Evaluations for Sustainable Development:
Experiences and Learning

The Editors

Aniruddha Brahmachari, is a social sector professional with 20 years experience in Programme Management, Monitoring and Evaluation. Mr. Brahmachari, presently leads the Monitoring, Evaluation and Learning (MEL) division of OXFAM India and is a member of the coordination committee of MEL, OXFAM International. He has worked with LANDESA, IDE, UNDP India, and CARE India. He also headed Social Research division in AC Nielsen ORG-MARG Nepal and spearheaded a wide range of research projects for AC Nielsen ORG MARG India. He obtained M. Phil in Planning and Development from Indian Institute of Technology, Bombay and Masters in Planning and Development from University of Poona and Master of Business Management.

Dr. Rashmi Agrawal is working as Director in NILERD (formerly IAMR). She is a trained professional in Monitoring and Evaluation from IPDET, Carleton University in Canada and a member of IDEAS Board and a founder member of Community of Evaluator South Asia. She is involved in conducting evaluations at various levels.

Samik Ghosh, has 10 years of experience working in social development sector and presently associated with OXFAM India for coordinating Monitoring Evaluation Learning work for its country programme and is engaged as team member of OXFAM International responsible data programme policy. He is qualified and awarded with a degree Master of Arts in Social Work (specialization in Community Health Management and Psychiatric Social Work) from Banaras Hindu University in India. He has worked on issues of Maternal and Child Health Nutrition, Participatory planning, Social Research, Monitoring and Evaluation, Learning for development processes in India and other countries in South Asia. He has previously worked with Government of India supported National Rural Health Mission programme, Academy for Educational Development Inc. for USAID/A2Z India Project and FXB India Suraksha.

Nabesh Bohidar, works with CARE India in Delhi. He has over 17 years of experience in Research and Management. He has authored a number of publications; on women's empowerment, rights of vulnerable populations, Cross-border Migration, Governance, Sports and on HIV and AIDS. He obtained M.Phil from Jawaharlal Nehru University, New Delhi, India.

Evaluations for Sustainable Development:
Experiences and Learning

— *Editors* —

Aniruddha Brahmachari

Dr. Rashmi Agrawal

Samik Ghosh

Nabesh Bohidar

2015

Daya Publishing House®

A Division of

Astral International Pvt. Ltd.

New Delhi – 110 002

OXFAM India
4th and 5th Floor, Shriram Bharatiya Kala Kendra
1, Copernicus Marg,
New Delhi-110 001 (India)
Phone: +91-11-46538000, Ext. 168, Fax: +91-11-46538099
E-mail: aniruddha@oxfamindia.org
Website: www.oxfamindia.org

Published by	:	**Daya Publishing House®** *A Division of* **Astral International Pvt. Ltd.** 4760-61/23,. Ansari Road, Darya Ganj, New Delhi - 110 002 Phone: 011-4354 9197, 2327 8134 E-mail: info@astralint.com Website: www.astralint.com
Laser Typesetting	:	**Classic Computer Services** Delhi - 110 035
Printed at	:	**Replica Press**

PRINTED IN INDIA

Acknowledgement

As editors of **"Evaluations for Sustainable Development: Experiences and Learning"** on behalf of NITI Aayog, Government of India we wish to acknowledge each of authors included in this book. We express our gratitude to the editorial committee who undertook initiative to support enormous process. We were very fortunate to have received inputs from dispersed expertise across various sections of book.

Editorial committee sincerely acknowledges support and encouragement of NITI Aayog and NILERD, New Delhi, India.

We were extended excellent cooperation around timeframe by Mr. Anil Mittal of Astral International Pvt. Ltd.

In all edited books there is a challenge to hold the central theme. Our authors did a tremendous effort to turn their papers to incorporate addressing the central theme of this book. In fact, some of them received these suggestions as almost an afterthought: our sincere thanks to them.

As editors we are grateful to our families and our mentors.

Aniruddha Brahmachari

Dr. Rashmi Agrawal

Samik Ghosh

Nabesh Bohidar

नीति आयोग

National Institute for Transforming India

Government of India

Madan Mohan

Adviser, PEO

Message

I am happy that the Programme Evaluation Organization (PEO) of NITI Aayog and National Institute of Labour Economics Research and Development have jointly taken the initiative to organize Evaluation Week from 19th to 23rd January 2015 to mark the start of the International Evaluation Year (2015). Evaluation is an important tool for testing the efficacy and effectiveness of policies and programmes in their contribution to development goals and improving their performance, thereby contributing to good governance. India has a long history of evaluations of development programmes, commencing way back in 1952 with setting up of PEO under Planning Commission. So far about 300 studies have been conducted by PEO which have provided valuable inputs for policy formulation and mid-course corrections.

I congratulate NILERD for this excellent and timely initiative. I understand that a number of evaluators from national and international agencies have provided technical and logistic support in the organization of this event. I appreciate their contribution.

Madan Mohan

राष्ट्रीय श्रम अर्थशास्त्र अनुसंधान
एवं विकास संस्थान

NITI AAYOG
Government of India

National Institute of Labour Economics Research and Development
(Formerly Institute of Applied Manpower Research)

Dr. Yogesh Suri
Director General

Message

I am glad to note that in recognition of importance of evaluation and advocating the same towards evidence based policy making, the year 2015 has been declared as the International Year of Evaluation (EvalYear) at the Third International Conference on National Evaluation Capacities held in Brazil in October, 2013. In this context NITI Aayog and its autonomous institution viz. National Institute of Labour Economics Research and Development (NILERD) (formerly Institute of Applied Manpower Research), alongwith co-partners are organizing the EvalWeek during January 19-23, 2015. A number of events are proposed to be held during the week commencing with a seminar on Evaluations for Good Governance at India Habitat Centre, New Delhi on January 19, 2015. This is expected to be followed by many similar events across the globe during 2015.

Government of India has replaced Planning Commission with a new Institution viz. NITI Aayog i.e. the National Institution for Transforming India. NITI Aayog is expected to be the think tank of the Government of India providing strategic and technical inputs across the spectrum with dissemination of best practices, infusion of new policy ideas and issue based support. Amongst the major objectives of setting up NITI Aayog is to actively monitor and evaluate the implementation of programmes

सैक्टर ए-7, नरेला संस्थागत क्षेत्र, दिल्ली-110040 (भारत), Sector A-7, Narela Institutional Area, Delhi 110 040 (India), Tel: 27787215-17, Fax: 91-11-27783467, Email: iamrindia@nic.in, Web: http://www.iamrindia.gov.in

and initiatives including identification of needed resources so as to strengthen the probability of success and scope of delivery.

I would like to thank various partners of NITI Aayog and NILERD in organizing this event and bringing out this compendium of articles covering good practices in Evaluation. I am sure that deliberations of these events and articles of the compendium would provide useful strategic inputs in formulation and restructuring of various schemes and programmes at state, regional, national and international levels.

I wish the events all success.

Dr. Yogesh Suri

No. PC/MoS(P)/(IC)/ _01_ /VIP/2015

राव इन्द्रजीत सिंह
RAO INDERJIT SINGH

राज्य मंत्री (स्वतंत्र प्रभार)
योजना मंत्रालय और
रक्षा राज्य मंत्री
भारत सरकार, नई दिल्ली—110001
Minister of State (Independent Charge) for
Ministry of Planning and
Minister of State for Defence
Government of India, New Delhi-110001

सत्यमेव जयते

Message

The year 2014 has been a landmark year for India. Pursuant to General Elections, a number of new initiatives have been launched by the Government of India such as the Make in India campaign, Prime Minister's Jan Dhan Yojana, Swachch Bharat Abhiyan and Digital India. Good Governance is the central theme of all these initiatives and evaluation remains a pertinent part of good governance.

I am happy to note that the year 2015 has been declared as the International Year of Evaluation and NITI Aayog alongwith National Institute of Labour Economics Research and Development and other co-partners are organizing Evaluation Week (EvalWeek) in Delhi during January 19-23, 2015. I appreciate the compendium of articles being released on this occasion and hope it would provide useful inputs to policy makers, researchers, academicians and other stakeholders.

I congratulate the organisers and wish the event all success

(Inderjit Singh)

Contents

Section V:
INNOVATION IN EVALUATION

EVALUATION POLICY
AND ETHICS

Chapter 1

Policy Support to Evaluations for Good Governance and Sustainable Development

Rashmi Agrawal[1] and B.V.L.N. Rao[2]

[1]NILERD (formerly IAMR), Delhi
E-mail: rashmi_agrawal56@rediffmail.com
[2]ISS Retired, Free Lance Evaluation Professional

ABSTRACT

Good governance implies effective delivery of economic and social services to the community for inclusive and sustainable development.Evaluation is an important tool in this process. For evaluations to be effective and usable they should be backed by National Evaluation Policy. The paper outlines the national and international efforts in framing and advocating evaluation policies. It observes that in spite of a long history of evaluation there is no explicit evaluation policy in India and empasises need for a flexible well articulated evaluation policy with details of its components. It suggests three main segments that should get attention in the policy which include the conceptual clarity of the evaluation process with details of terms used in evaluation, standards, procedures, dessimination and feed back loops for evaluations conducted and an action plan that specifies the roles and responsibilities. Institutional arrangements should also be a part of policy document.

Keywords: *Good governance, Sustainable development and evaluations, National evaluation policy and its segments.*

Good Governance

Good governance stems from various structures put in place to implement decisions to give effective national developmental priorities and policies. Good governance does not depend on the size of these structures but on effective functioning of those structures through mutual coordination, cooperation and relevant communication within and outside the government. Governance involves formal and informal actors, government functionaries among the former and the civil society organizations among the later. Minimum government and maximum governance is often stressed by the Indian Prime Minister as key to optimum results in developmental interventions. Good governance also implies inclusive development which is essential for sustainable development, as embodied in the sustainable development goals to be adopted by UN.

Excellent planning may fail in the absence of good governance. A number of projects that were planned well failed due to lack of commitment, willingness and overarching guidance in their execution. This phenomenon can be observed across nations and across projects and programmes. While it is important to make good decisions it is also important how decisions are made. This reflects upon the process that is involved in making decisions.

The Indian government is emphsising the need for good governance embodying the cardinal principle that even the weakest and the most vulnerable sections of society have an equal stake in charting the country's growth. The model of good governance ensures that even the government should be answerable to even an ordinary citizen, hailing from any part of India.

Evaluation as Tool for Good Governance

Good governance comprises of a number of components. Australian Public Service (2007) indicated them as strong leadership, culture and communication, appropriate governance committee structures, clear accountability mechanisms, working effectively across organisational boundaries, comprehensive risk management, compliance and assurance systems, strategic planning, performance monitoring and evaluation, flexible and evolving principles-based systems. The Eleventh Five Year Plan of India indicated that the essential components of good governance are constitutionally protected right of people to elect government at all levels in a fair manner, accountability and transparency of government at all levels, effectiveness and efficiency in delivery of social and economic public services, empowerment of lower level government, rule of law and fairness and inclusiveness in the system. The critical element in good governance is thus effective and efficient implementation of social and economic policies and programmes that ensure inclusiveness in the system.

It may be concluded that performance monitoring and evaluation, structures for ensuring accountability are crucial aids to good governance. As stated by Heider,C governance and evaluation are interdependent. While good governmance creates an enabling environment for evaluations, evaluations contribute to good governance.

Indian System of Evaluation and Governance

India is a multi-party federal democracy with three main levels of governance. The three levels are Union, States and Local bodies. The Union system includes various ministries and a number of apex organizations while State system has State departments and local governments include rural and urban structures with elected representatives. To have checks and balances there is an independent judiciary and other constitutional organizations like Election Commission, Comptroller and Auditor General, Central Vigilence Commission, etc. Media and civil society organizations play their part in public affairs and their influence on issues relating to governance has increased considerably in recent times.

Inclusive social and economic development is a priority of the governance in the country and policies and programmes in furtherance of this goal are conceived, planned and implemented at various levels of governance. There are a number of major development programmes of national importance such as those on Education for All, Health for All, National Rural Employment Guarantee, Mid Day Meal, National Rural Drinking Water Program and so on which are given special attention by the Union government. Implementation of these national programmes and schemes as well as the state specific programmes conceived at that level are generally implemented through the states and local governments while national coordination is done at central level.

To ensure effective and efficient delivery of social and economic services to the citizens, the key element of good governance, the existence of a robust system of monitoring and evaluation (M and E) is essential. The twin processes of M and E in the country have as long a history as development planning itself. The very First Five Year Plan (1951-56) suggested that how new policies and programmes are being received and what effects they have are questions that arise at every stage in the implementation of a plan of development and recommended that systematic evaluation should become a normal administrative practice in all branches of public activity. The Second Plan (1956-61) clarified that the purpose of evaluation was to assess whether the programme was succeeding in its fundamental objectives.

M and E has been assigned a key role in meeting the objectives of developmental planning and specific institutional mechanisms therefore were put in place. The Programme Evaluation Organisation (PEO) was set up in the Planning Commission of India as an independent body in 1952, initially for assessing the work of national extension and community development programme. It now operates through seven Regional Offices and eight field offices. These institutions have been taking up evaluation studies on the development programmes of various Ministries in the central government, usually at the request of the latter and on occasion on the initiatives of the Planning Commission itself. Similar institutions came up within the planning departments of most State governments during the sixties and early seventies. A Development Evaluation Advisory Committee (DEAC) advises PEO on prioritization of areas of research, methodologies to be adopted, establishment of linkages between PEO and various evaluation research organizations and academic institutions and follow-up action on evaluation results. The structure of Planning Commission and its mandate is in the process of complete change to make it relevant to the people following a bottom up approach with decentaralisation of decision making of planning and implementation. The Commission has been abolished and a new institution has come up as National Institution for Transforming India (NITI Aayog). The new institution will give due attention to monitoring and evaluation of development interventions to make them effective in outreach and impact.

In recent years, evaluation of development projects of the government is increasingly being done by independent research organizations, independent evaluators and others in the voluntary sector, for better transparency and objectivity.

Although PEO has been responsible to organise evaluations of various developmental interventions, unfortunately monitoring was never in its domain. This function has generally rested with the implementing agencies. The Planning Commission till now had been taking note of the progress of various programmes while allocating resources through annual and five-year plans. The finance departments also monitor progress through performance budgeting. In recent years the office of the Prime Minister has been excersising the monitoring function very closely through Results Framework Documents (RFD). Informally considerable monitoring is done by civil society organizations, media and community.

Policy Support to Evaluation

It is seen from the foregoing that India has evolved institutional structures and mechanisms for using evaluations as a tool for good governance. However, it cannot be concluded that evaluations have invariably contributed to optimizing the returns from developmental programmes in terms of impacts. There are several reasons, the fundamental one being the absence of a coherent National Evaluation Policy (NEP). A policy can provide a direction in which all the partners can move to make interventions effective and results-oriented. M and E has come a long way from traditional monitoring of utilization of funds allocated and coverage of physical targets fixed to assessment of impact on the lives of people. In spite of the existence of an evaluation system in India that has deep roots, there is no national evaluation policy. Absence of well-articulated NEP leads to pitfalls like inadequate awareness, sensitivity and appreciation of the importance of evaluation as a means to optimize performance results leading to enhanced standards of living of the people. There is also a problem of ownership of evaluations conducted and response of the management to evaluations. There are instances where evaluations are done in a routine manner sometimes by a number of organizations at the same time for the same program and neither the sponsors nor the evaluators own the evaluations as evaluators feel no responsibility for a follow up once report is submitted and sponsors meet the objective of organizing evaluation and the matter ends there leading to low levels of utilization of evaluation findings in the development effort. The structure of evaluation in India is fractured and there is lack of a common unifying environment of evaluation culture. While the demand for evaluations is emerging from various corners there is totally inadequate attention to capacity development on the supply side which puts a question mark on the quality of evaluations. An explicit evaluation policy can address most of these deficiencies.

Developing Countries and Evaluation Policies: A Review

While developing countries generally have some sort of monitoring and evaluation system, explicit NEPs are yet to be adopted in most countries. Eval Partners in collaboration with UN Women and other organizations have recently brought out a web-publication on 'how to integrate gender equality and social equity issues into national evaluation policies and sys-tems (NEPSs) that are being implemented in an increasing number of developing countries around the world, with the aim of making them gender-responsive. The document is intended for all of the different public and private-sector agencies involved in the design, implementation and use of evaluations of development policies and programmes as well as organizations concerned about ensur-ing that evaluations address issues such as gender equality, social equity and human rights.' (2014). According to this study, there are 16 developing countries across the globe – 6 with well-established NEPs, 6 where NEPs are in an evolving stage and 4 where they are in an early stage.

Table 1.1: Status of NEPs in Developing Countries

Country	Income Level	Is there a Single Document Creating and Defining NEPs	Stage of Development	Coverage
Benin	Low	Yes	Early stage	Whole-of-government system
Bhutan	Low	No	Early stage	Whole-of-government system
Chile	High	No	Well established	Whole-of-government system
Colombia	Upper middle	No	Well established	Whole-of-government system
Costa Rica	Upper middle	Yes	Evolving	Whole-of-government system
Ethiopia	Low	Yes	Well established	Only covers certain sectors
India	Lower middle	Yes	Evolving	Whole-of-government system
Kenya	Low	No	Evolving	Only covers certain sectors
Kyrgyz Republic	Low	Yes	Early stage	Whole-of-government system
Malaysia	Upper middle	No; a number of legislative decrees	Well established	Whole-of-government system
Mexico	Upper middle	No	Well established	Whole-of-government system
Morocco	Lower middle	Yes	Early stage	Whole-of-government system
Nepal	Low	Yes	Evolving	Whole-of-government system
South Africa	Upper middle	Yes	Well established	Whole-of-government system
Sri Lanka	Upper middle	No	Evolving	Whole-of-government system
Uganda	Low	Yes	Evolving	Whole-of-government system

Source: Bamberger M, Segone M, Reddy S.2014. *National evaluation policies for sustainable and equitable development -How to integrate gender equality and social equity in national evaluation policies and system.* P13, www.mymande.org/selected-books

According to the above table The Republic of South Africa has a well established NEP. The policy draws its legal basis from the Policy Framework of the Government-wide Monitoring and Evaluation System approved by the Cabinet. Such legal backing enhances the chances of effective enforcement of the policy. In Malaysia evaluation is integrated as a key factor into the performance planning. It has a unique multi-dimensional evaluation capacity development approach which is built upon tripartite collaboration of the public, private and civil society sectors (Stolyarenko, retrieved 2015).

Community of Evaluators (CoE)- South Asia have taken an intitative of creating a task team on Enabling Environment on Evaluations with the purpose of developing a Framework for Evaluation Policy in South Asia. The document has been finalized (CoE, 2014) which aims at achieving the following:

☆ Contributing to the development in South-Asian region by improving the quality and usability of evaluations at country level through well-defined conceptual frameworks and guiding principles and standardized protocols.

☆ Promoting standards in evaluations across South Asian Region, building on the key principles and good practices of evaluation across the globe

☆ Enhancing credibility and transparency in evaluations for better outcomes and impacts

☆ Emphasizing the importance of skill development, providing guidelines about capabilities and skills needed and specifying the roles of various stakeholders in evaluations for better governance and effectiveness of developmental interventions

☆ Sensitizing evaluators, decision makers, program implementers and managers about importance of evaluation.

The Framework also proposes a template that can be utilsed to develop national policies on evaluations.

Towards NEP in India

Absence of a clear evaluation policy for public investment and development programmes may lead to evaluation practice that is ad hoc, at times merely symbolic, aimed at only justifying and expanding what the policy makers and programme planners anyway want to do, not usable because of poor quality and, often, not leading to significant returns from investments in evaluations themselves. Therefore, in India, while there have been on occasion quality evaluations that have generated eminently useful results which have contributed to performance enhancement, policy modifications and improvement in various programmes, there have also been several of indifferent quality and merely symbolic. It is thus essential that evaluation practice is backed by a clearly enunciated policy and more importantly a commitment on the part of evaluation commissioners, mangers and evaluators to implement the policy in letter and spirit.This section discusses some of the essential aspects that such NEP should focus upon.

The NEP should have three parts – the first relates to the conceptual clarity of the evaluation process with details of terms used in evaluation so that entire process is demystified.The objectives and rationale of framing a policy should also be brought out. The second part should contain specifications about standards, procedures, dissemination and feed back loops for evaluations conducted. It should also mention about the ethics and ownership for utilization of evaluations., standardization of skills and credentials of evaluators should also have a place in this section. The third component is an action plan that specifies the roles and responsibilities of various actors in implementation of the policy and how they should go about implementation of the policy. Institutional arrangements and development of evaluation capabilities should also be a part of this segment of policy document.

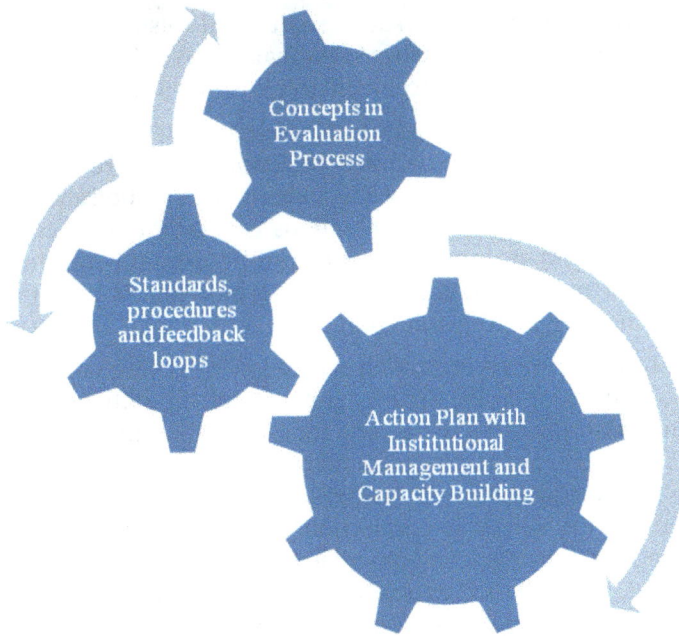

Figure 1.1: Evolving Evaluation Policy.

Purposeful evaluations are possible only in an environment that is aware of and recognizes the importance of evaluations and their contribution towards improving programme results. Policy makers, including legislators, should be aware and be convinced of the essentiality of periodic evaluations in evolving policies and managing programmes. It is equally necessary that evaluation practitioners realize how important is the objectivity of the results of their investigations for policy makers. The community stake holders also need to know that the evaluations conducted provide objective tooth and are indeed utilized by the policy makers in shaping the future of the country. Sensitization of all the sections of the evaluation community including people at large is the key to success of evaluations. The sensitization has to come from awareness generation workshops at policy levels, mapping of best practices and their demonstration and involvement of community in development and evaluations.

While evaluation is becoming more familiar to orgasnisations, many donot appreciate or visualize the ultimate utility of the evaluation findings (Leviton, 2013). Sensitization should lead to better appreciation of the benefits of evaluation findings, ownership of evaluations and better utilization of evaluation results at all levels including organizations. The communities, government and other agencies involved with evaluations should feel that such excercises are for the betterment of the society as a whole and a sense of belongingness of the programmes as well as evaluations will emerge. For this purpose, the policy should emphasize the need for involvement of people in the development planning, implementation, monitoring and evaluations through innovative participatory methods. It is also necessary that the policy mentions about the earmarking the resources, in doin g so it should also provide for the crunch situation indicating practical approaches to evaluations. A judious choice of practical evaluations can optimize the resources available (Agrawal, 2013).

Even though the evaluation activity has increased by leaps and bounds, there is no explicit institutional mechanism to track the action taken on recommendations of various evaluations. There is also lack of adequate communication and dissemination plan for evaluations with the result evaluations generally remain confined to the sponsoring agency and do not come into public domain limiting action on recommendations.The policy should elaborate this issue and provide for creation of a tracking mechanism- with dissemination plan.

One of the factors effecting utilization of evaluations is the quality of evaluations which in turn depends upon the skills of those involved with evaluations. While there had been a quantum jump in demand for evaluations, the supply of qualified evaluators has not kept pace and lags behind considerably. The Working Group on Strengthening Monitoring and Evaluation System for the Social Sector Development Schemes in the country set up in connection with the Tenth Five year plan also emphasized need for capacity development in this area not only in the public sector organizations but also in the civil society organizations and research institutions. Evaluation theory and practice is constantly evolving and newer methodologies are coming up in response to freshly emerging needs. Further the bottom up approaches of evaluations more in vogue these days also need to develop capacities at community level.The evaluation policy of the country should specify the modus operndi and institutional mechanism with time line for developing capacities at various levels. Institute of Applied Manpower Research (IAMR) has taken an initiative in this direction by starting a diploma course and other short term programmes in M and E for national and international participants. The M and E as a separate discipline is yet to emerge in India as well as in other countries. A critical aspect in this field is the need to establish competency standards. Some such documents are available (IDEAS, Canadian Evaluation Society) which can be perused for adaptation as per local needs.

Policy should stipulate that every development intervention should compulsorily include a set of indicators of performance. A model set of indicators can be developed drawing upon international and national experience and documentation for major programmes so that monitoring and evaluation of such programmes could be standardized at various levels and places. For instance some effort had been initiated in developing the RFD. The new programmes like 'Jan Dhan Yojana' that have great potential for outreach and community development needs to be monitored from the very beginning and periodic evaluation should be a part of the scheme to make it effective and effecient. There is a need to identify indicators for monitoring of such ambitious schemes. The data so captured would also facilitate evaluations of such programmes in the future.

As the programmes and schemes vary considerably in their objectives, size, budgets, implementations plans and their relative importance in the country, the NEP should have in built flexibility to take care of such variations to some extent in terms of which programmes and schemes must be provided with funds for evaluations, their frequency and timing, the methodology to be adopted and so on.

The importance of evaluations can be demonstrated no better than by evaluating the evaluations themselves and bringing out the benefits that have accrued to the policies and programmes by acting on the evaluation results. Such review would also highlight lessons to be learnt for future evaluations as well as best practices. This exercise can lead to further methodological advances as well as help in triangulating results of different evaluations conducted on the same subject. The knowledge so generated can be shared across nationally and internationally.

Framing of NEP by itself is not an end. A demonstrated commitment on the part of all stakeholders for its implementation is sine qua non. The experience of some of the countries like Sri Lanka shows that till such commitment comes from the legislators or other statutory bodies, the whole effort in framing a policy can be fruitless. Similarly some states like Karnataka in India formulated the policy but it is yet to be accepted and implemented. A better way would be to have a National Policy of Evaluation which could be adapted at various levels as per local needs, traditions and terminology.

A coherent and well-articulated NEP incorporating issues of gender equality and social equity backed by a firm commitment to implement the policy would take evaluation practice to a plane higher than now, thereby establishing it as an effective tool for better delivery of economic and social services for sustainable development by promoting good governance.

References

Agrawal. R.,2013, Resource Crunch, Evaluations, and Mindset, Development Evaluation in Times of Turbulence, *Dealing with Crises That Endanger Our Future*, edited by Ray C. Rist, Marie-Helene Boily Frederick R. Martin, The World Bank, pp. 69-77.

Bamberger M, Segone M, Reddy S.2014. *National evaluation policies for sustainable and equitable development -How to integrate gender equality and social equity in national evaluation policies and systems.on line available at December, 30th, 2014* www.mymande.org/selected-books

Case-study #6: National evaluation policy in malasiya, Parliamentarians Forum for Development Evaluation, Katerina Stolyarenko, Independent Consultant, *betterevaluation.org/resources/./national_evaluation_policy_in_malaysia, Jan.6, 2015*

Commonwealth of Australia, Australian Public Service Commission, 2007, *Building Better Governance,* www.apsc.gov.au/__data/assets/pdf_file/0010/./ bettergovernance.pdf

Draft proposal for The National Evaluation Policy of The Government of Sri Lanka, Sri Lanka Evaluation Association, June 2006

Three months' Diploma in Monitoring and Evaluation, Institute of Applied Manpower Reasearch, Government of India details available on www.iamrindia.gov.in

Framework for Evaluation Policy for South Asia, Community of Evaluators- South Asia. *2014* under print

IDEAS Competencies for Development Evaluation Evaluators, Managers, and Commissioners was adopted by the IDEAS Membership on 07 February 2012. Available on www.ideas-int.org/

GOI, Planning Commission, 1951, First Five Year Plan (1951-56)

GOI, Planning Commission, 1956, Second Five Year Plan, (1956-61)

GOI, Planning Commission, 2007, Eleventh Five Year Plan, (2007-12)

Haider. C. 2011, A Conceptual Framework for Developing Evaluation Capacities; Building on good Practice. P. 85-110. In Influencing Change, *Building Evaluation Capacity to Strengthen Governance,* edited by Ray C. Rist, Marie-Helene Boily Frederick R. Martin, The World Bank,

Leviton,L.C., 2013, Some Underexamined Aspects of Evaluation Capacity Building reprint in *American Journal of Evaluation*, Vol.35(1) p 90-94.

The Canadian Evaluation Society. Competencies for Canadian Evaluation Practice. Available on www.evaluationcanada.ca/

The Presidency Republic of South Africa, Department: Performance Monitoring and Evaluation, National Evaluation Policy Framework, 23 November 2011 (Final).

The Working Group on Strengthening Monitoring and Evaluation System for the Social Sector Development Schemes, Tenth Five Year Plan (2002-07)

Chapter 2

What Prevents Creating Evaluation Environment?

I.C. Awasthi[1] and G.P. Joshi[2]

[1]Professor, Giri Institute of Development Studies, Lucknow, U.P.
E-mail: icawasthi@gmail.com
[2]Deputy Director, [1]NILERD (formerly IAMR), Delhi
E-mail: joshigp1960@gmail.com

ABSTRACT

Clearly, there is disconnect between the information gathering, monitoring and measuring impacts. The paper argues that there is a need for monitoring and evaluation framework in place in every program that eventually aids to improve the design and delivery of projects, programmes and policies. There ought to be visibly clear linkages between evaluation findings and resource allocation in order to narrow down the hiatus between the intents and outcomes of the programs.

Massive public investments are being made on development programmes and obviously governments and other stakeholders want to know how well and to what extent the delivery mechanism is achieving the desired goals or intents of policies. There is, therefore, a need for credible evaluation framework in place in every project or programme that eventually aids to improve the design and delivery of projects, programmes and policies and it must move beyond an emphasis on inputs and outputs to a greater focus on outcomes and impacts or results. There is a need for building institutional capacity, developing capabilities and competencies and strong demand for ownership of an evaluation system in order to spreading evaluation culture as part of economic reforms.

I. Introduction

The outcomes or ends of development need to be expressed in terms of human welfare and the real ends can be expressed in terms of providing health, education and employment and these are the insignia of progress. Growth is indeed important and without growth resources cannot be generated for investment in social infrastructure, particularly primary health and primary education, and physical infrastructure. But growth has to be inclusive with shared benefits. In order to ensure that growth to be broad based and inclusive there is a need to implement effective public policies and strategies that bring more people in to mainstream, promote the increased participation of poor men and women in economic decision-making, reduce poverty, disparities and promote equitable access to the benefits of growth. It is now increasingly realized that development goals needs to be expressed in explicit and real terms so that the fruits of development are widely shared. This ought to be important policy objective and rightly so where a large segment of population is marginalisd, socially and economically excluded. Where social security is an integral part of governance, it becomes an effective strategy in carrying out both protective and promotional security measures. It would, therefore be imperative necessity to empower them through protective social security. Indeed provision of employment is the most powerful instrument to provide social security.

Successive five year plans have made huge public investments in social and economic sectors that have resulted in enormous benefits to the society and people in terms of output, employment and incomes. Indeed, employment generation has been one of the principal concerns of development planning in the country. However, growth of employment has generally been slower than the growth of labour force. While the annual growth of gross domestic product (GDP) grew over time, the employment did not increase correspondingly. Most of the GDP growth was thus derived from productivity growth rather than increase in employment. As a result, employment elasticity of GDP growth declined. Evidence shows that employment growth has been sluggish in recent times and sometimes termed as jobless growth. Most of the employment is being generated in the unorganized sector of economy which is by and large precarious in nature.

It is well known that the quality of employment in the informal sector is low, wages and working conditions are precarious. It is important from the policy point of view to make the employment productive through specific interventions that could help to develop more dynamic and better-protected informal sector and its progressive integration in to society. This would require improvement of the productive capacity through improvement in training and technology, establishing an appropriate regulatory framework including appropriate forms of protection and regulation.

Conventionally, governance structures have been characterised by rule-based approaches that primarily focused on process regulation, compliance with centrally prescribed standards and rules. Performance has been thus judged not by the results or outcomes but by compliance with processes and inputs. This has severely undermined the performance of the development projects or programmes. However, in a changed economic milieu, the performance is evaluated by results or outcomes not by the outputs; the fallacy of mere outputs has been seriously questioned in evaluation literature (Kusek and Rist 2004, Linda and Rist 2009).

II. Why Evaluation?

Historically, evaluation grew out of the auditing tradition in 19th century in Britain that essentially focused on verifiable and dependable financial records necessitated from the growing commercial activities. Then governments and organizations moved from auditing and inspection to an emphasis on impact and development evaluation sub-discipline drew heavily on scientific and social research methods. Organisation for Economic Co-operation and Development (OECD) played important role in advancing development of evaluation. OECD's development assistance committee (DAC) network on evaluation primarily aims at increasing the effectiveness of international development programmes by supporting robust, informed and evaluation quality standard (Furubo et.al 2002). Evaluation enables to improve the design and delivery of projects, programmes and policies that in turn help how well and to what extent the delivery mechanism achieving the desired goals or intents of policies. OECD countries have matured in monitoring and evaluation system and have been instrumental in adopting and spreading the evaluation culture. In OECD countries and other developed countries there have been strong internal and external pressures that led to development of evaluation culture. This culture has been spreading gradually in policy domain in many developing countries.

There has been huge growth of professional evaluation associations in developing and developed countries aimed at capacity development in development evaluation. Within the World Bank an independent evaluation group (IEG) was set up in 1995 which has undertaken large numbers of country programme evaluation. Yet another development in international organization for evaluation is the International Development Evaluation Association (IDEAS) started in 2001 with a view to strengthening the evaluation capacity.

III. Appraisal of Evaluation in India

Despite the fact that evaluation organization is in place since India's first five year plan, no substantive and credible evaluation work has been done nor has evaluation culture been extended in the policy domain in real sense.

A large number of centrally sponsored and central sector schemes are implemented through different Ministries across the country. With enormous diversity in the implementation hierarchy across space, it has become all the more important to have information about the physical and financial details of a project or a programme in order to monitor the progress. The Eleventh Five Year Plan underscored the deficiencies in the existing accounting system for the Plan schemes and its inability to support informed planning, budgeting and effective monitoring of these schemes (Planning Commission 2008).

Government of India has launched 13 flagship programmes with a view to creating and strengthening infrastructure, promoting employment and livelihoods and improving health, sanitation etc. Massive public investments have been made in flagship a programme which is of the order of Rs.1900 billion in 2012-13. When huge public investment is being made obviously, governments and other stakeholders need to know how well and to what extent the delivery mechanism is achieving the desired goals or intents of policies from huge investments. Indeed, without any credible monitoring and evaluation system in place the efficacy and effectiveness of these programmes will remain largely unknown.

However, no road map with clear actions for development evaluation has been prepared in meeting challenges of development. Some of the problems noted include the following:

(i) Most of the projects or programmes has inbuilt monitoring system in terms of physical and financial targets during the currency of project or programme. However, targeting approach has number of lacunae as it simply tries to track the progress in terms of physical and financial targets. In other words such a mechanism addresses the inputs, activities and outputs but rarely it focuses on outcome and impacts. It misses the result based evaluation.

(ii) Huge information is collected in every project or programme but the information is hardly used for analytical purposes with a view to aiding policy or assessing the impacts. Even the evaluation studies sparsely use evaluation framework and robust methodology that often results lack of analytical rigor and comprehension. Such results have little or no usability for assessing the impact of a project or programme or even helping for mid-course corrections.

(iii) Many a times no quantifiable indicators are developed towards achieving an outcome or impact (result). Outcome cannot be measured directly therefore there is a need for translating in to set of quantifiable indicators.

(iv) In majority cases evaluation studies do not have base line data as a result it becomes extremely difficult to assess the real impacts of the project or programme.

(v) Purpose of evaluation and benefits thereof are rarely defined. Relation between monitoring and evaluation is often missed or not properly understood. Also, what type of evaluation approaches are to be employed is not clear (formative, summative or prospective, for instance).

(vi) Evaluation hardly takes place in policy domain and there is lack of evaluation culture in the public spending. Also, there are very few champions that influences the policy makers in order to give evaluation due importance in the economic establishments.

(vii) There are very few independent evaluation studies taken up by the credible individuals or institutions and most of the evaluations lack ethics and professionalism. In this situation it is difficult to foresee the utility of such evaluation studies.

(viii) The log frame is hardly used scientifically in order to outline the defined and realistic objectives and assumptions that describe how the interventions are designed to work. Log-frame hierarchy lacks relationship and links among inputs, activities, outputs, programme objectives and outcome. It does not appear that flagship programmes are assessed in the log frame hierarchy which is indispensable in order to judge the efficacy of the programme.

(ix) It is also important to take on a concurrent evaluation of the project or the programme in order to understand the activities with a view to achieving the programme objectives. The concurrent or formative evaluation helps understand the bottlenecks and impediments in the process. It examines the planning and implementation process and suggests corrective measures of the programme. This involves discussions with all the stakeholders in order to make the programme effective through qualitative assessment supported with quantitative analysis. This activity must enhance the overall decision- making process. However, the missing link in the implementation process is the concurrent monitoring that has been noted even in the Mahatma Gandhi National Rural Employment Guarantee Act which has a good management information system in place (Ambasta 2009). It has been noted that monitoring and evaluation functions in a compartmentalized manner rather than articulating relationships between the two. For instance, the Ministry of Rural Development has an in-house monitoring wing but the evaluation studies are carried out by other agencies.

(x) It is often observed that there is a weak connection or linkage between implementation framework and outcome framework within the perspective of theory of change. This arises due to lack of clarity on the indicators and assumptions of the log frame. It is also noticed that there is a divergence between the project or programme data and that of large survey data that often makes difficult to aid policy (Integrated Child Development Scheme and National Family Health Survey, for example). It is therefore necessary to have a common design and data collection framework in order to have comparability between the different sets of data sources. The survey data can serve as base line in order to establish initial conditions against which the effects of a completed project or programme can be compared. Clearly, connecting output, outcome and survey data appears to be missing.

(xi) No explicit results chain (or theory of change) appears to underlie the choice of indicators in flagship programmes. This is despite the fact that Government of India has promulgated performance monitoring and evaluation guidelines that task government departments to design and use results frameworks for all major programmes (Awasthi 2013).

IV. Way Forward

There is a growing need for effective and efficient utilisation of public money through continuous monitoring process in order to measure the progress. Precisely, with this in view, every project and programme has some monitoring process in place in order to track the progress. However, there has been severe disconnection between implementation framework with that of outcome framework or results. Our monitoring and evaluation set up has been obsessed with providing information on implementation as opposed to broader development effectiveness issues and heavily preoccupied with "are we doing the thing right" rather questioning "are we doing the right thing".

There is disconnect between the information gathering, monitoring and measuring impacts in most of the projects or programmes. Even having elaborate information collection system and monitoring mechanism in place does not necessarily lead to impact evaluation of most of the social programmes. Report of the Evaluation Gap Working Group (2006) aptly described "most of these resources are directed toward monitoring the use of funds, deploying and managing personnel, and producing outputs and services. By contrast, relatively little is spent to rigorously assess whether programs are having the desired impact beyond what would have occurred without them". There is greater demand for effective and efficient utilization of public money with clear impacts and governments are under enormous pressure to demonstrate results both on account of their survival and greater value to public money.

There is, therefore, a need for monitoring and evaluation framework in place in every project or programme that eventually aids to improve the design and delivery of projects, programmes and policies and it must move beyond an emphasis on outputs to a greater focus on outcomes and impacts or results. There ought to be visibly clear linkages between evaluation findings and resource allocation in order to narrow down the hiatus between the intents and outcomes of the programmes. We must clearly learn the lesson that what driving development evaluation in many developing countries with clear results and also to unlearn from our own experiences that after six decades of evaluation organization in place nothing has moved towards result based evaluation. Obviously, there is a need for building institutional capacity, developing capabilities and competencies and strong demand for ownership of an evaluation system in order to spreading evaluation culture as part of economic reforms.

References

Ambasta, Pramathesh. 2009. Programming NREGS to Succeed. *The Hindu.* October, 30. *m.thehindu.com/.../lead/programming-nregs-to-succeed/article41154.ece (accessed in December, 2014).*

Awasthi, I.C. 2013. Monitoring Public Investment: Lessons from India. *Evaluation Connections. News Letter of European Evaluation Society.* June, Prague 4, Czech Republic.

Center for Global Development.2006. *Report of the Evaluation Gap Working Group. When Will We Ever Learn? Improving Lives through Impact Evaluation.* Washington, D.C.

Furubo, Jan-Eric and Rolf Sandahl (Eds.). 2002. *International Atlas of Evaluation.* New Brunswick, NJ: Transaction Publishers.

Kusek, Jody Zall and Ray C. Rist. 2004. *Ten Steps to a Result-based Monitoring and Evaluation System.* The World Bank, DC.

Linda, G. Morra Imas and Ray C. Rist. 2009. *The Road to Results: Designing and Conducting Effective Development Evaluation*. The World Bank, Washington DC.

Planning Commission. Government of India. 2008. *Eleventh Five Year Plan* (2007-12). Vol. I, Chapter 3.

Chapter 3

Ethical Challenges for Evaluation in India

Alok Srivastava

Director-CMS Social,
Centre for Media Studies (CMS)
New Delhi
E-mail: alok@cmsindia.org

ABSTRACT

The paper based on the author's first-hand experience of the ethical challenges faced in India during carrying out social research and evaluation, in particular. It highlights the complexity of ensuring ethical norms in a country like India due to its diverse socio-cultural and multi-lingual population. In absence of any well-defined ethical guidelines for non-clinical evaluation or even research in India, it is mostly left to the choice of the evaluators to opt for an ethical review of their research protocols prior to initiating the evaluation. The paper highlights various aspects of ethical challenges, right from selection criteria followed for selection of an agency/individual for a sponsored evaluation/research to carrying out the assignment and the follow up with regard to utilization of evaluation findings. The paper concludes with practical solutions to overcome the ethical challenges faced while carrying out evaluation in India and may be adapted in other developing economies as well.

Keywords: Ethics, Challenges, Social, Evaluation, Research.

Background

The relevance and importance of practicing ethical norms is increasing day by day in the field of development evaluation or even in social research for that matter. However in a country like India, conducting ethically and at the same time scientifically rigour evaluation is a challenge in itself. Why? India is a country with more than 1.2 billion population constituting about 1/6th of world's population; around two-third of India is below 35 yrs age [Census of India 2011]. India is a predominantly agrarian country with more than 70 per cent Indians living in 638,000 villages. Rest live in more than 5,100 towns and over 380 urban agglomerations. Around 60 per cent workforce engaged in agriculture and allied activities but growth rate in agriculture sector is around 4 per cent. Officially around one-fourth live below poverty line of USD 1 but some estimate it to be as high as 45 per cent. To make the scenario more complex, India is a religiously, culturally diverse multi-lingual society. More than 18 major languages combined with some 1652 languages and dialects are being spoken in India. The challenge becomes more prominent because of low literacy rate. As per Census 2011, literacy rate is around 74 per cent; even lesser among female- 65 per cent than male-82 per cent. As a proverb has it, in India "Every two miles the water changes, every four miles the speech [James, Robert, 1995]. With such a socio-culturally diverse population, designing a uniformly acceptable ethically robust evaluation with human subjects is a challenge in India.

The paper elaborates on key challenges in conducting an evaluation, particularly primary data based ones, and where engaging with the community is a must. But before discussing the ethical challenges in conducting evaluation in India, it is important to understand, what ethics is and what is meant by human subjects in evaluation or research, in broader context?

What is Ethics?

When most people think of ethics (or morals), they think of rules for distinguishing between right and wrong, such as the Golden Rule ("Do unto others as you would have them do unto you"), a code of professional conduct like the Hippocratic Oath ("First of all, do no harm"), a religious creed like the Ten Commandments ("Thou Shall not kill."), or a wise aphorisms like the sayings of Confucius. This is the most common way of defining "ethics": norms for conduct that distinguish between acceptable and unacceptable behaviour [David 2011]. It is pertinent to mention that right from the childhood the grooming takes place on the ethical and morally correct behaviour. In other words, we learn from childhood at home, at school, in religious places, or in other social settings. Undoubtedly, in fast growing professional world of research, relevance and importance of practicing ethical norms is very critical as it ensures objectivity, promotes truth and knowledge and ensures lesser occurrence of error.

Human Subjects

The next question that arises is what is meant by human subject? The United States Department of Health and Human Services (DHHS) defines a human research subject as a living individual about whom a research investigator (whether a professional or a student) obtains data through 1) intervention or interaction with the individual, or 2) identifiable private information (32 CFR 219.102.f) [Lim 1990].

As defined by DHHS regulations:

☆ "Intervention"- physical procedures by which data is gathered and the manipulation of the subject and/or their environment for research purposes [45 CFR 46.102(f)].

☆ "Interaction" is communication or interpersonal contact between investigator and subject [45 CFR 46.102(f)].

☆ "Private Information"- information about behaviour that occurs in a context in which an individual can reasonably expect that no observation or recording is taking place, and information which has been provided for specific purposes by an individual and which the individual can reasonably expect will not be made public [45 CFR 46.102(f)]

☆ "Identifiable information" means specific information that can be used to identify an individual.

Thus, human beings irrespective of gender, age group, ethnic group and socio-economic status, individually or in group, considered as a 'subject' for research are identified as human subjects for social science research. However, research involving human subjects categorized in special categories such as minors, pregnant women, differently-abled, prisoners become ethically more sensitive. While on one hand, research involving human participants must not violate any universally applicable ethical standards, on the other hand, a researcher needs to consider local cultural values when it comes to the application of the ethical principles to individual autonomy and informed consent [ICMR 2006].

The paper primarily focuses on human subjects as a whole and look into challenges from ethical perspective. Important ethical issues include voluntary participation and informed consent, anonymity and confidentiality, and accountability in terms of the accuracy of analysis and reporting [Patrick 2010]. However, many a time ethical discussions usually remain detached or marginalized from discussion of research projects. In fact many researchers consider this aspect of research as an afterthought [Sharlene, Patricia, 2006].

But before we look in to these challenges, the question arises, why at all should we practice ethical norms in evaluation or social research with human subjects? The answers are many but simple. It prohibits immoral approach towards information/data collection. Further, restricts misrepresentation of information/data and restricts researchers from being biased. Also, to an extent, emotional conflicts of surveyed population are addressed properly. On researchers' part, accountability of researchers towards the community gets ensured and last but not the least, institutions/organizations more likely to fund research projects can trust the quality and integrity of research.

The basic demand of ethical norms is to respect human dignity and privacy; take special precautions with vulnerable population; and make efforts to ensure utilization of evaluation findings *i.e.* follow-up with donors/implementing agencies.

Ethical Practices and Challenges in India

As far as practicing ethical norms in evaluation, including social research in India is concerned, without hesitation one can say that in India no well-laid ethical guidelines are in place; but practiced more on a case to case basis. Ethical review of social research proposals and protocols are yet to be institutionalized. In India, Institutional Review Board (IRB) on ethics for non-clinical research is few, almost non-existent. Many universities in India have duly-constituted ethics committee but their review is limited to research by their faculty and students and not to research done outside the University purview.

In 1999, Ethical Guidelines for Social Science Research in Health was framed by the National Committee for Ethics in Social Science Research in Health (NCESSRH) [Centre for Enquiry into Health and Allied Themes]. It would not be wrong to say that non-clinical health evaluation/research to an extent follow some basics of ethical clearances but in most of the cases it is more of a voluntary choice and less as a pre-requisite for initiating a research study. Most often practiced ethical norm in India is to take 'consent' of the respondents and that too mostly as part of studies related to some socially sensitive issues such as HIV/AIDS, reproductive/sexual health topics or for collection of blood samples. Most of the time, the consent is verbal in nature due to poor/low literacy status of respondents. Even here, all risks and benefits are not detailed out while reading out the consent statement. It is often argued that by agreeing to participate in the survey, it is presumed that the consent for participation has been given by the respondent!! As a matter of fact in majority of the cases, it is out of respect, particularly in rural India, that a person agrees to participate in the survey rather than by understanding and absorbing the objectives of the study or the pros and cons of their participation. As a result they might not share facts but give politically correct answers to questions, which make them uncomfortable.

Problems of confidentiality are also sharper in qualitative research. Quantitative evaluators can often deal with confidentiality issues through the sampling process, and through technical safeguards when the data is analyzed. We may decide to guarantee the privacy of disclosures. But there are two difficult questions. First, do all research participants have equal rights to privacy; and second, do a commitment to privacy cover all circumstances? (Shaw2003)

It is often said in context of social research in India is that formal ethical review of research protocols are undertaken only when the institution or the Principal Investigator(s) is keen to publish some research papers/articles in journals of repute. If the research is meant to suffice donor's need only or to strengthen one's credentials from business aspect, then hardly efforts are made to get the research protocol reviewed by and institutional ethical review board.

The ethical clearance to an extent ensures that the research team will strictly abide by the method and approach suggested in the duly-approved research protocol by the ethical review body. The reality as observed is that at ground level *i.e.* during data and information collection, while interacting with community or just before selecting the human subject or respondent (interviewee), the researchers/study team may sometimes revise/reselect the sample *i.e.* goes for convenience sampling. Since no system is in place nor it is mandatory to do post-check of whether the sample selection was followed as suggested in the research protocol, the challenge that arises is 'how to control the deviation from the proposed approach.' This may further lead to biased findings being reported. In other words, deviation from originally proposed sample design should be considered unethical because the findings on important indicators, for instance, reach of a social welfare programmes or initiatives may be misleading.

In areas affected by natural calamities or even civic unrest (riots), research team assessing the benefits of relief work has to be always careful in ensuring that the respondents/beneficiaries' sentiments are not hurt. For instance, the purpose of the research may be to assess the reach and impact of the benefits and services provided during the calamity or unrest but the recipients should not be subjected to any kind of embarrassment or harassment because of the enquiries made. Ethically it would be wrong to put them in any kind of agony and distress by posing questions related to the pre and post situation. How to overcome the situation is an ethical challenge for research team?

Financially compensating the participating human subjects is an often discussed ethical concern. Reimbursement of costs to participants in research stipulate that it is generally appropriate to reimburse the costs to participants of taking part in research, including costs such as travel, accommodation and parking. Sometimes participants may also be paid for time involved. However, payment that is disproportionate to the time involved, or any other inducement that is likely to encourage participants to take risks, is ethically unacceptable [National Statement on Ethical Conduct in Human Research, 2007]. Particularly requesting local target population (mostly daily wage earners, marginalized and deprived community) to participate in evaluation process as respondent or participant of Participatory Monitoring and Evaluation exercise, such as using Participatory Learning and Action (PLA) techniques, many a times make them lose their wage for a day or two as they will have to remain absent from work. Even if we ignore the implications on research cost/budget, the question that arises-Is compensating by paying token money at prevailing wage rate ethically right? It has been observed that in many situations it has led to commotion within the community to be one of the participants. Another important aspect that needs to be cautious about is that findings may also get biased as they may not give adverse comments about the programme/project/process, as they are being 'paid' to participate ....or may even give too adverse comments without any substantial evidence if they feel the researcher is keen to hear on those lines. A lot of debate has taken place on this issue of reimbursing the participants for their participation but no conclusive decision has been arrived at.

Global economic and cultural power relations also raise ethical challenges in cases of outsiders studying relatively poor people in other places, potentially contributing to a strong asymmetry in the researcher-informant relationship, and in worst case contributing to insufficient care for the interests of the informants. Another ethical challenge connected to studying the community – without interfering in - a potentially ethically problematic issue such as surrogacy is the risk of implicitly contributing to legitimating problematic activities and practices. Further the role of a social researcher limits the degree of interfering in the field, - where the approach of the researcher to maintain non-interfering position or not is an ethical challenge.

Another critical concern is conflict of interest, which is becoming a major ethical challenge not only in the growing competitive business world but evaluation/social research too. Many agencies, national as well as internationals' scope of business activities or area of operation is expanding. As a result, while they are funding some projects and at the same time are into implementation of similar nature of programmes and projects and simultaneously are in to the business of providing consultancy services for evaluation and assessment of such kind of programmes and projects. At one point of time, they invite bids/proposals for evaluating their programmes/projects and at another point of time, they compete with the similar or even same evaluating agencies to bid for evaluation of any outside programme/project. Members of evaluation society also bid for evaluation projects and are also employees/consultant in funding/donor agencies, which invite proposals from institutions/agencies providing evaluation consultancy, thereby having access to technical and financial bids of the evaluation institutions/agencies.

Selection criteria followed of bidding institution/individual also have an ethical concern attached to it. By and large Quality and Cost Based Selection (QCBS) approach is commonly practiced but it has been noticed in India that many a times if 3S *i.e.* Size, Spread and Selection of Sample for the research is not pre-defined in the Terms of Reference of the bidding document, then selection of lowest financial bidder is ethically wrong. This method is only appropriate for selecting consultants for assignments of a standard or routine nature (audits, engineering design of noncomplex works, and so forth) where well established practices and standards exist [MoHRD 2009]. As such selection of research agency based on the lowest financial bid, commonly called as L1 may lead to compromising with the quality of evaluation findings because the sample suggested may not be representative.

Evaluation findings should be put in public domain or not is another issue of debate and has ethical perspective as well. It becomes important to discuss in public domain the larger issue concerning society. In case the sponsor of the study is not willing to share the unfavourable findings in public then how to overcome the challenge without compromising with client-relationship ethics and at the same time responsibility towards larger audience. Moreover, ethically is it the responsibility of evaluating agency/consultant to do follow-up with the funding agency of the assignment to know whether evaluation findings have been shared with target population and other stakeholders or not and what action have been taken on the recommendations/suggestions made in the report.

Conclusion

Some of the discussed issues related to ethical practices in research involving human subjects are country specific but by and large stand valid for all developing economies of the world. It is important to ensure that research is ethically sound in terms of being objective, unbiased and uninfluenced by any external or internal factors.

Four Pointers for Action for ensuring ethically acceptable evaluation and research are:

1. Evaluators' community should ensure establishment of Institutional Ethical Review Boards (IERB) accredited at the national level by the government with well-represented board of experienced development evaluators and social researchers. It should be mandatory for all social research bodies to seek approval from IERB for each study undertaken by them and an unique IERB Approval number must be put on all research/evaluation documents including research tools and report.

2. Develop an ethical guideline manual for individual as well as institutional development evaluating agencies/institutions. A pre-requisite for all such institutions and individuals to accept the laid down instruction in the guidelines.

3. Introduce/Promote course in research ethics; training in research ethics to help research/evaluation/donor community.

4. To create awareness among community at large about importance of practicing ethical norms in research. This will help to build public support for evaluation, as they can trust the quality and integrity of the exercise, which in turn improves the quality of evaluation and research, as a whole.

References

American Evaluation Association, 1994, *Guiding principles for evaluators: New Directions for Evaluation, 66, pp. 19-26.*

Census 2011, Office of the Registrar General and Census Commissioner India, Ministry of Home Affairs, Government of India, Available online at www.censusindia.gov.in.

Corti, L., 2000, *Progress and Problems of Preserving and Providing Access to Qualitative Data for Social Research—The International Picture of an Emerging Culture*, Forum Quality Social Researching, 1(3). Available online:http://www.qualitative-research.net/fqs-texte/3-00/3-00corti-e.htm#g7

Indian Council of Medical Research, New Delhi, 2006, *Ethical Guidelines for Biomedical Research on Human Subject.*

James H. and Robert L. Worden, 1995, *India: A Country Study.* Washington: GPO for the Library of Congress.

Lim, 1990, *What is Human Subjects Research?* University of Texas at Austin

Ministry of Human Resource Development (MoHRD), Government of India, 2009, *Procurement Manual, Technical Education, Quality Improvement Programme*

National Committee for Ethics in Social Science Research in Health (NCESSRH), accessed at Centre for Enquiry into Health and Allied Themes' website http://www.cehat.org/publications/members.html

National Statement on Ethical Conduct in Human Research, 2007. Available online at: www.nhmrc.gov.au/publications/synopses

Patrick D., 2010 *Ethical Dilemmas in Sampling*, Journal of Social Work Values and Ethics.

Resnik DB, 2011, *What is Ethics in Research and Why is it Important?*

Sharlene N. H. and Patricia L., 2006, *The Practice of Qualitative Research*, Sage Publication

Shaw, I.F., 2003. Ethics in Qualitative Research and Evaluation. *Journal of Social Work*, 3(1), pp. 9-29.

Chapter 4

Programme Evaluation and Evaluation Policy of Karnataka 1969 to 2014

Brijesh Kumar Dikshit

Chief Evaluation Officer,
Karnataka Evaluation Authority,
Bangalore, Karnataka
E-mail: keapd20111@gmail.com and bkdceo@gmail.com

ABSTRACT

Evaluation is an important tool for effective programme implementation. Karnataka recognized the need for evaluation as early as 1969. It formed the first Evaluation Policy in 2000. The Policy was revised in 2011 incorporating lessons learnt in implementing the Evaluation Policy of 2000 and the Karnataka Evaluation Authority formed. Karnataka is the only State to have an Authority committed to Evaluation.

Keywords: *Evaluation policy, Types of evaluation, Process of evaluation.*

Introduction

In independent India, the formation and implementation of five year plans has been the means to achieve the goals of development and growth. Elaborate and sophisticated techniques have been employed to formulate programmes and schemes to achieve these goals. But schemes and projects sometimes fail to achieve what they were intended for. This calls for an in depth analysis of the concept and the process of implementation, which diagnoses the weaknesses and shortcomings, and suggests corrective measures. This process is the process of evaluation of the programme/scheme.

Dawn of the Concept of Evaluation in Karnataka

It was as early as 1969 that the State of Karnataka realized with experience gained since 1951 that each investment in programmes and schemes needed much attention since poor project preparation and analysis invariably resulted in unsound investment, delays in execution, high costs and low yields. The need for analyzing every programme and scheme using scientific basis and modern techniques was recognized, and evaluation of programmes and schemes was commenced with publication of "*Principles of Evaluation – A Manual*" in 1969 by the Directorate of Evaluation and Manpower Planning and Social Welfare Department of the then Government of Mysore. The publication contained the definition, types and tenets of Evaluation; Statistical techniques to be utilized in Evaluation, and ended with a chapter on Report writing.

Evaluation in the Recent Past in Karnataka

Karnataka was one of the first States of the Country to come out with an "*Evaluation Policy*" in the year 2000. The policy provided on outlay earmarked for undertaking quick evaluation studies by external agencies including Directorate of Economics and Statistics for every major scheme, programme and project in the annual plan. It formally provided for:

a) All schemes with a budget outlay of more than Rs. One crore should be evaluated by an external agency at once in five years,

b) Up to 1 per cent of the project cost, subject to a ceiling of Rs. 5 lakhs earmarked for evaluation,

c) If any scheme was to continue beyond a plan period, it ought to be justified through an evaluation, and,

d) The outcomes of evaluation should be used for improving programme design and delivery.

There were two Committees, one chaired by the Secretary in charge of various departments of the Government and the other chaired by the Chief Secretary, Government of Karnataka who were to monitor the implementation of the Evaluation Policy.

Lessons Learnt by Implementing the Year 2000 Evaluation Policy

After the implementation of the year 2000 Evaluation Policy, introspection was done about a decade later by the Planning, Programme Monitoring and Statistics department of the State. The following shortcomings in implementing the Evaluation Policy were documented:

1. Absence of well formulated programmes with clear verifiable outcomes and benchmarks against which evaluation was to be done.

2. The staff in the Evaluation division of planning department did not have adequate technical competency to get evaluations done.

3. Evaluation reports were not of very inspiring quality as external agencies doing evaluation did not have the expected capacity.

4. Since the department getting evaluation done was also paying for the evaluation study, there was a conflict of interest.

5. There was no system of reward and punishment for evaluation. As a result there was no motivation for departments to get evaluation done. At times, evaluations were done in a monotonous way with imperfect vendor selection.

6. Many evaluation study findings had not been meaningfully used.

Formation of Karnataka Evaluation Policy and Evaluation Authority

It is against this history of Evaluation and the lessons learnt during the implementation of year 2000 Evaluation Policy, with a view to overcome the shortcomings and lacunae in evaluations of the past and to have a professional, unbiased and independent body with the responsibilities of carrying out evaluations that the Government of Karnataka sanctioned the "*Karnataka State Evaluation Policy and Karnataka Evaluation Authority*" in the year 2011. The Karnataka Evaluation Authority is a society registered under the Karnataka Societies Act 1960.It receives funds for evaluation studies from the Government of Karnataka and also some of its departments. It has been exempted from paying Income Tax by the Income Tax department.

Main Points of Evaluation Policy of 2011

The mission of the Evaluation Policy of 2011 is to have a "*transparent, effective and efficient policy of evaluation of its development policies and programmes*". It provides for prescribing standards of evaluation and strives to enhance the technical capacities within the departments of the Government and independent evaluation agencies to undertake and utilize evaluation outputs. This policy includes all departments, urban and rural local bodies, corporations and State owned industries. The policy envisages two broad types of Evaluation, namely:

1. External Evaluation

This is where the Karnataka Evaluation Authority (KEA) initiates Evaluation studies that are paid by its own funds. When a scheme is taken up for external evaluation by KEA, it cannot get evaluated by the concerned department. The line department is duty bound to furnish all information to KEA in time, and extend all help in the carrying out the evaluation.

2. Internal Evaluation

An evaluation taken up by the line departments from their own resources is referred to as Internal Evaluation. In order to ensure that there is no conflict of interest, it is mandatory for the concerned department to take advice from the KEA on the Terms of Reference(ToR) of the evaluation study, data collection tools, vendor outsourcing, all other technical matters and to follow the rules and procedures prescribed by KEA.

The Evaluation Policy has made provision that any evaluation can be handed over to KEA at any stage.

The Results Framework Document (RFD) includes a weightage for getting Evaluation done of the major programmes of all departments, corporations, industries and urban/rural local bodies. The progress made in RFD is reviewed by Special Committees which includes the Principal Secretaries and heads of departments. This has provided a great fillip to Internal Evaluation.

Functions of the Karnataka Evaluation Authority

The functions of KEA are:

1. To supervise, facilitate, build capacity and handhold departments for effective Planning, Monitoring and fine tuning the policies, programmes and schemes.

2. To undertake or commission training, consultancy, advocacy activity to further goals of effective and meaningful scheme formulation, Monitoring and Evaluation.

3. To keep record of all Terms of Reference(ToR) of Evaluation Studies, data collection tools, evaluation reports and to follow up utilization of evaluation outputs.

4. To formulate rules and procedures for selection of agency for Evaluation and publication of a training manual on Evaluation.

5. To disseminate the findings of evaluation studies.

Process Followed by KEA in External Evaluation

The process followed by KEA in External Evaluation is:

1. The theme/programme/scheme intended to be evaluated is decided. This may be upon the reference of the scheme by the General Body of the KEA or by resolution of the Governing Body of KEA.

2. The line department related to the title of Evaluation study is requested to provide all background material of the programme/scheme selected for evaluation. This includes the objectives of the scheme, the target group it is expected to serve, the physical and financial achievements, findings of any previous evaluation studies and the follow up action taken on them.

3. The office of KEA prepares a Terms of Reference (ToR) of the Evaluation study and sends it to the line departments for vetting, improvement and approval.

4. This approved ToR is placed before the Technical Committee of KEA. This Committee is chaired by the Principal Secretary, Planning, and has Professors of Indian Institute of Management, Bangalore, Institute of Social and Economic Change, Fiscal Policy Institute, National Sample Survey

Organization etc. as its members. The ToR is discussed and approved with or without changes.

5. After the approval of ToR, the Evaluation study is outsourced to Consultant Organization. 39 Consultant Organizations are empanelled with KEA. The outsourcing is done in accordance with the Karnataka Transparency in Public Procurement Act 1999 and Rules made there under in 2000.

6. The Consultant Organization given the Evaluation study prepares and presents to the Technical Committee of KEA a Work Plan detailing how they would go about with the study. The work plan includes sample size, sample selection method, data collection tools, how data will be analyzed etc. Only after the Technical Committee approves the Work Plan the study proceeds.

7. After data has been collected, the Consultant Organization prepares a draft report and sends it to KEA. An internal committee of KEA goes through it and, if they find it *prima facie* acceptable and in accordance with the ToR and standards of Evaluation, send it to an Internal Assessor. These Internal Assessors are government officers and academicians with fairly long experience. They go through the draft reports and send back the report with comments for improvement. These are sent to the Consultant Organization for incorporation/changes.

8. After the comments of the Internal Assessor are complied with, the final draft report is put before the Technical Committee of the KEA. The line department to which the study is related is always invited to that meeting. After the line department agrees that the report is acceptable, the final draft report is discussed and, along with inputs of the line departments and members of the Technical Committee, the report is accepted after all the inputs are incorporated and complied with.

9. The final report is received by KEA. Copies of it are sent to the line departments and the officers of Planning department. One copy of the report is put up on the website of KEA.

10. The progress of Evaluation Studies is reviewed by the General Body of KEA which is headed by the Chief Secretary to the Government of Karnataka and also in the State Karnataka Twenty Point Programme review.

11. The Development Commissioner of the State convenes a meeting of line department Officers, Consultant Organizations and KEA to discuss the findings and recommendations of evaluation studies. The meeting discusses what recommendations are to be acted upon, how they are to be acted upon and also the follow up of action resolved to be taken.

Process Followed in Internal Evaluation

The process followed in Internal Evaluation is the same as that of External Evaluation, except that the scheme selection is done by the line departments and cost of Evaluation is borne by them. The line departments decide the Consultant Organization in consultation with KEA.

Progress Made by the KEA

Karnataka Evaluation Authority has prepared and published the following Manuals relating to Evaluation Studies:

1. The Manual for Empanelment of Consultants and Internal Assessors.
2. The Manual for Output Grading of Evaluation Studies.
3. The technical Manual for Evaluation.

The achievement of KEA in conducting Evaluation Studies is as follows:

Sl.No.	Work Type	Achievement in			
		2011-12	*2012-13*	*2013-14*	*2014-15 Upto Dec. 2014*
1.	Approval of Terms of Reference of Evaluation Studies	0	7	19	35
2	Outsourcing External Studies	2	4	15	23
3	Completing Evaluation Studies	0	2	6	23

Conclusion

Karnataka has recognized the need and importance of Programme Evaluation since 1969. It formulated an Evaluation Policy in the year 2000.In implementing that it learnt that a professional approach to evaluation, eliminating conflict of interest and a structured process and format of evaluation is necessary if Evaluation has to succeed. Evaluation will be infructous if there is no mechanism by which the findings and recommendations of evaluation studies are acted upon. The new Evaluation Policy of 2011 and the formation of Karnataka Evaluation Authority seeks to set right the anomalies noticed hitherto and provide environment for a professional, unbiased and effective evaluation.

References

Directorate of Evaluation and Manpower 1969. Principles of Evaluation- A Manual: Government of Mysore Publication.

Government of Karnataka order number IFS 42 EVF (1) 99 dated 17th November 2000.

Government of Karnataka order number PD 8 EVN (2) 2011 dated 11th July 2011.

Government of Karnataka order number PD 23 PSD 2014 dated 11th April 2014.

Chapter 5

An Evaluation of the System of Evaluation for Developmental Programmes in India

Manoj Kumar Mishra

Programme Evaluation Organisation (PEO), Planning Commission
Regional Evaluation Office, Planning Commission,
Kolkata, West Bengal
E-mail: manojmishra37@gmail.com

ABSTRACT

Poor evaluability of developmental programmes in India is the main challenge for the evaluation. Outcomes of most of the programmes are not adequately defined in terms of quantitative verifiable variables. Multiple programmes with overlapping objectives and unavailability of data are the other constraints. However, Programme Evaluation Organisation (PEO), and other institutions have brought about excellent evaluation results. Most of the flagships programmes have been evaluated by PEO. This paper will discuss in brief the findings which have major policy implications. The issues left unanswered or have become relevant over period of time and which should be taken into consideration for any future evaluations will also be highlighted.

Government of India has taken some good initiative in public sector management like necessitating the preparation of Outcome Budget and Result Framework Document by all departments. Some social sector programmes require collection of information on baseline and regular progress. These initiatives are very supportive for the evaluation works.

This paper also highlights the need for institutional capacity building of the evaluation organization at the centre and the state level.

Keywords: *Flagship programmes, PEO, CAG, Performance audit, Evaluability, Impact, Outcome, indicators, and Baseline.*

Introduction

In India evaluation is an integral part of the programme implementation and the concerned line ministries/departments are entrusted with this function. They normally outsource the evaluation to some agencies. There is an in built mechanism introduced in each of the flagship programmes for earmarking certain percentage of total funds annually for monitoring and evaluation. The Programme Evaluation Organisation (PEO) undertakes evaluation studies as per the requirements of the various divisions of the Planning Commission and the ministries/departments of Government of India. Comptroller and Auditor General of India (CAG) also conduct 'Performance Audit'. These are independent assessment or examination of the extent to which a programme/scheme operates economically, efficiently and effectively. In the states the responsibilities of evaluation falls on the State Evaluation Organisations (SEOs). These organizations are required to evaluate the adequacy of the planning, implementation methods and impact of development programmes.

The focus of development evaluation has been shifting over the years. Traditionally evaluation was linked with the implementation. It did not provide the government with an understanding of success and failure of the policies/schemes. It was designed to address the question 'Did they do it?' Modern evaluation answers the question 'So what?'. The aim of evaluation now is to determine the relevance and fulfillment of objectives, efficiency, impact and sustainability of the policy/ programmes.

Quality evaluation of various programmes and projects would not only bring improvement in public sector performance but would also address broad range of issues relating to economy, efficiency, sustainability and relevance of public funding and development interventions.

Evaluability Assessment

One needs to know where he is going, why he is going and how he will know when he gets there. The intended outcomes should be disaggregated sufficiently and the process and outcome indicators should be appropriately crafted. It should provide simple and reliable means to measure achievement that should answer the question "what would we expect to see as verifiable evidence of achievement of the outcomes or impact".[1] However, in most cases the programme outcomes are not properly defined in verifiable quantitative terms. Evaluators have to face methodological challenges in measuring the changes overtime. Some examples of such objectives are as follow.

Programme	Objectives
Swachh Bharat Mission	1. To bring about behavioural change in people regarding sanitary practices. 2. Generate awareness among the citizens about sanitation and its linkage with public health.
Mid Day Meal Scheme	1. To address hunger in schools by serving hot cooked meal. 2. To improve nutritional status of children.
Hill Area Developmen Programme	1. Eco preservation and eco restoration with focus on sustainable use of bio diversity 2. Ensuring community participation in the design and implementation of strategies for conservation of bio-diversity and sustainable livelihoods

However some programmes do have SMART[2] indicators. Here is one such example.

Programme	Objectives
National Health Mission	1. Reduce MMR to 1/1000 live births 2. Reduce IMR to 25/1000 live births 3. Reduce TFR to 2.1

In this direction Government of India has taken some welcome initiatives in the public sector management. The Outcome Budgeting has been started since 2005 which would provide projected outcome and quantifiable variable. Another initiative was taken in 2009-10 when government introduced a system of Result Framework Document for all the departments/ministries.

Secondly, many programmes are designed and implemented as standalone schemes, though the target groups and the outcome variables are common in many cases. These multiple programmes and their structure of implementation are unrelated and sometimes conflicting too. For example, Minimum Support Price and Targeted Public Distribution System (TPDS) are interrelated programmes. But, these two programems are implemented differently by different departments.

Thirdly, most of the programmes are universal in nature, hence construction of meaningful counterfactuals for evaluation purpose is very difficult. Moreover, the goals of many programmes are also overlapping and the objectives have been changing overtime. For example, the objective of Mid Day Meal (MDM) programme has been shifted from universalization of education to improvement of nutrition and health of the children. Construction of any logical model (theory of change) is very difficult in such cases.

Fourthly, specific and sufficient information (data) are not available for evaluation work. Many programmes are not being systematically monitored and the data on progress are not being collected on outcome indicators. Most of the flagship programmes of Government of India *e.g.* Integrated Child Development Services (ICDS), Total Sanitation Campaign (TSC), require collection of baseline information and online monitoring. However information is collected mostly on physical progress with little information on intended outcomes. Moreover, the genuineness of the data is doubted. For example, the Report of the Evaluation Study on TSC conducted by PEO has noted the divergence in the ratio of achievement to targets available at TSC website and the data available with the Census of India. PEO conducts sample survey to collect data (primary/secondary) from the field for the evaluation studies. This resulted in the delay in the completion of evaluation work. Due to undefined objectives and spatial diversity in the implementation, there is lack of clarity about what data should be monitored, collected and analyzed. However, in recent year changes are apparent in the public sector management in general and implementation of flagship schemes in particular.

In spite of all these constraints, institutions involved in evaluation (mainly PEO) have come up with wonderful evaluation results. Some of these are given below

Food Security

The Targeted Public Distribution System (TPDS), which envisaged that below poverty line (BPL) families would get a certain quantity of food grains at subsidized price, was introduced in 1997.

A study conducted by PEO found that 58 per cent of the subsidized food grains do not reach the intended BPL families. It has estimated that 36 per cent of the budgetary subsidy on food was siphoned off the supply chain and 21 per cent reach to APL families. The study observed targeting error, prevalence of ghost cards and unidentified households. The study further suggested that for one rupee worth of income transfers to poor the government spends Rs. 3.65 indicating that only one rupee budgetary subsidy is worth only 27 paise to poor.

Issues for future evaluation: Shift from universal Public Distribution System (PDS) to TPDS was based on the notion that subsidized grain would reach those who actually needed it. However, the implementation is plagued by targeting error. The evaluation of mechanism of identifying BPL is beyond the scope of TPDS. As stated earlier, the impact of TPDS cannot be understood without studying govt. support price and procurement policy including the functioning of Food Corporation of India (FCI).

Education

Mid Day Meal (MDM was launched in 1995 with twin objectives *i.e.* universalisation of primary education (by improving enrolment, attendance and retention) and improvement of nutritional status of children in general and those belonging to disadvantaged sections in particular. The scheme was revised in 2002, 2004, 2006 and 2007. There was qualitative shift in the focus of scheme in 2006 from education to nutrition and health. As of now the sole objective of the programme is to improve nutrition and health of elementary children and improvement of attendance and concentration of poor and disadvantaged children.

CAG in its performance audit of MDM observed that no system had been established to assess the outcomes of the scheme in terms of a well defined parameter (nutritional status of children). It has also found that the data on enrollment collected from the states were inconsistent with the data maintained by the centre (M/o HRD) which indicate unreliable data capture. The social audit conducted by Government of Andhara Pradesh with MV Foundation in 2008, also observed that there was a tendency to inflate figures of attendance due to corruption and other reasons.

On the basis of its field observation CAG remarked, "The MDM is being implemented with primary purpose for providing one daily meal without any link to education, nutrition and health objectives."

PEO which conduction evaluation study in 2006-07 found that the scheme has no significant impact on enrolment in majority of the sample schools (except in Andhra Pradesh and Madhya Pradesh).

CAG has also estimated loss of 11-30 hours per week by the teachers due to MDM. PEO in its report observed deviation of education time by students for washing utensils.

Sarva Shiksha Abhiyan (SSA) was conceived at the end of Ninth Five Year Plan to improve accessibility, quality and reduce gender and social gap in elementary education. PEO has conducted and an evaluation study on SSA in 2008. The study found that there had been good impact on enrolment. It has been able to reduce gender and social gap in case of enrolment.

However, CAG in its performance audit of 2006 observed mismanagement at various levels. The Annual Status of Education Report (ASER) prepared by Pratham shows little impact on the level of learning of the children.

Issues for future evaluation: In absence of any system of collection of information on nutritional status of children, assessment of the outcome of MDM is not possible.

As per ASER 2013 more than 96 per cent of the rural children in the age of 6-14 years are enrolled in schools. The out of schools children are confined to few pockets only. However large numbers of students is dropping outs in elementary education.

Any analysis on elementary education is not full without assessment of the role played by private schools. As per the ASER 2013 report, private school enrolment for rural children in the 6-14 years age group rose from 16.3 per cent in 2005 to 29 per cent in 2013.

Sanitation

The evaluation study conducted by PEO on TSC estimated that 73 per cent of rural households are practicing open defecation and 66 per cent of rural households are practicing this due to lack of toilets. However, it has also been observed that among the households having toilets 22 per cent are still resorting to open defecation. Lack of awareness and established age old practice have been cited as main reasons for this.

Furthermore, only 59 per cent toilets (provided through TSC) are both covered and have roof and 85 per cent toilets were of single pit. In many states low cost toilets had been adopted. For such toilets, a squatting plate having mosaic pan with inbuilt P-trap is placed over a pit. It can be covered with locally available materials like bamboo mat, jute, cloths etc. This is affordable by everyone and people can be easily motivated to accept, adopt and use such toilets. It was expected that the households would be induced to upgrade it due to behavioural changes. This strategy was quite successful in case of Intensive Sanitation Programme (ISP) in the Medinipur district of West Bengal which later became the base for the launch of TSC in 1999. But, this strategy was not effective at all India level. It was observed that 61 per cent households were not satisfied with TSC and the low cost latrine was the major cause of dissatisfaction.

There was another operational deficiency as only 46 per cent household reported adequate water for flushing.

Regarding institutional toilets, another evaluation study on ICDS by PEO found that 61.5 per cent anganwadis were without toilet. Anganwadis operating from private buildings was the main constraint in achieving the 100 percent target.

Information Education and Communication (IEC) was the most important component of the programme. Motivators are to be engaged at the village level for demand creation and taking up behavioural change communications. However, only 46 per cent gram panchayats have recruited motivators. In many states motivators have not been recruited at all.

However, the evaluation report highlights significant impact in the NGP awarded Gram Panchayats. People have reported remarkable decrease in open defecation and reduction in medical expenses due to improved sanitary conditions. As a result of these the average morbidity in NGP awarded GPs is lower than that of the other GPs.

Issues for future evaluation: The Nirmal Bharat Abhiyan (NBA) (effective since April 2012) or the latest Swachh Bharat Misssion (effective since 2nd October 2012) has the provision for Rs. 12000 incentive for individual household latrine. The incentive was also extended to indentified APL families besides the BPL families. Introduction of TSC in 1999 was an historical shift from high to low subsidy and supply driven to demand driven regime. Now the strategy has been again shifted to high subsidy for the household. There is need to compare the results of different approaches for sanitation overtime over time.

Nutrition and Health

With an aim to break the inter-generational cycle of malnutrition, reduction of morbidity and mortality caused by nutritional deficiency Government of India launched Integrated Child Development Services (ICDS) in 1975. PEO commissioned and evaluation study which was outsourced to National Council for Applied Economic Research (NCAER). The study estimated that only 31.1 per cent of the intended beneficiaries received supplementary nutrition out of total eligible children in the country.

National Rural Health Mission (NRHM) seeks to provide accessible, affordable and quality health care to rural populations, especially vulnerable and underserved population groups in the Country. PEO commissioned an evaluation on NRHM in 2011. The study reveals that Aante Natal Care (ANC) is still a grossly neglected area in maternal health care, as 22 per cent of the pregnant women in the sample states are not availing of ANC. The study also shows that 51 per cent of the mothers who delivered children in the last 5 years have not availed of any post natal care.

Issues for future evaluation: Both the programmes (ICDS and National Health Mission) have inbuilt mechanism which generate good amount of information on many indicators. However, the reliability of such information is questioned. PEO has just completed "A Test Check Study" to verify the reliability of such information. The findings of such study will be of great help in future evaluation. Such study should be conducted for other programmes too.

The evaluation studies conducted by different organizations are under criticism for their alleged inefficiency and limited utility. The evaluation studies commissioned by implementing departments are viewed as accommodative of the expectations of the departments. However, evaluation studies conducted by PEO have been able to pinpoint the success or failure of many programmes. But, due to the constraints, it has not been able to venture into multiple programme regularly. In most of the states there exist state level evaluation and monitoring committees headed by a minister or the chief secretary or the secretary of the concerned department. But the required human resource and technical facilities are not available in the SEOs. The officials posted in these organizations also lack exposure to the tools and techniques of development evaluations. Their works are more focused on monitoring than evaluation. There is need to strengthen these institutions as they have the potential to track the progress and feed this into the process of governing and decision making.

1. UNEG Handbook for Conducting Evaluation of Normative Work in the UN System
2. SMART: Specific, Measurable, Attainable, Relevant and Time bound

References

I. Books

Kumar Alok, Sqatting with Dignity, Sage Publication

II. Reports

1. Programme Evaluation Organisation; Planning Commission; Government of India: Performance Evaluation of Targeted Public Distribution System (TPDS)

2. Programme Evaluation Organisation; Planning Commission; Government of India:Evaluation Study on Total Sanitaion Campaign;

3. Programme Evaluation Organisation; Planning Commission; Government of India: Evaluation Report on SSA

4. Programme Evaluation Organisation; Planning Commission; Government of India:Evaluation Study of Performance Evaluation of Cooked Mid-Day Meal (CMDM)

5. Programme Evaluation Organisation; Planning Commission; Government of India:Evaluation Study on Integrated Child Development Scheme (ICDS)

6. Programme Evaluation Organisation; Planning Commission; Government of India: Evaluation Study on National Rural Health Mission (NRHM) in Seven States

7. Comptroller and Auditor General of India, 2006 Performance Audit of the Implementation of the SSA; Report No. 15

8. Comptroller and Auditor General of India, 2008 Performance Audit on National Programme for Nutritional Support to Primary Education

9. Pratham, India; 2014, Annual Status of Education Report (Rural) 2013

Chapter 6

A Policy Paper: Reformation and Rationalization of CHCs in Haryana

Ashish Gupta, Vivek Sharma and Pratyush Bishi

*Haryana State Health Resource Centre (HSHRC),
Panchkula, Haryana
E-mail: hshrcphpme@gmail.com, hshrcpkl@gmail.com*

ABSTRACT

The Community Health Centre is an important structure at its secondary position of public health care delivery system. The CHCs are responsible for clinical, preventive and promotive health services. It is also a referral and supervision centre for the PHCs and SCs in the periphery.

CHCs were established long back in health system in Haryana. Since then the population has increased many folds but the CHCs have not been rationalized or upgraded. Therefore, there are discrepancies in the availability of services, human resource, utilization and coverage of population as per Indian Public Health Standards (2012).

A Policy document Reformation and Rationalization of CHCs in Haryana based on a study conducted in CHCs of Haryana titled "Availability of Human Resource and Health Services and their utilization at CHCs of Haryana". The study has found out lacunas in various areas like inadequate HR, availability of essential health services, population coverage and its poor management.

This paper has recommended possible policy level solutions in terms of i) categorization of CHCs, ii) rationalization of population coverage, ii) public health cadre and iv) minimum assured health services to make CHCs effective and efficient in its functioning and maximally utilized with quality of services. The reformation and rationalization of CHCs has been also validated through GIS mapping.

Introduction

The Community Health Centre (CHC) has been envisaged for both primary and secondary health care services which cover rural and semi-urban population. The

CHC at its position as referral unit for PHCs and SCs plays an important role in contribution in achieving universal health care services. Location, geographical area of coverage and population covered under the CHCs are important determinants of access and utilization of these health institutions. The population coverage can be a major player on deciding on the appropriateness of the establishment of the CHCs and rationalization of the human resource.

The Indian Public Health Standards (IPHS) are the benchmarks which sets minimum standards for infrastructure, service availability, human resource and quality standards for different public health institution in India. IPH Standard (revised 2012) prescribes a minimum list of five services that are to be provided by the CHCs namely a) OPD and IPD services- General, Medicine, Surgery, Obstetrics and Gynaecology, Paediatrics, Dental and AYUSH services b) Emergency Services c) Maternal Health, New-born Care, Child Health and Family Planning Services d) Laboratory Services e) Various National Health Programs. As per the IPHS, CHCs are generally 30 indoor bedded hospitals with one operation theatre, labour room, X-ray, ECG and laboratory facility.

Functions of CHC

It is apparent from the IPH standard referred above that a Community Health Centre has multiple functions. It has been given the responsibility of clinical services, implementation health programmes, promotive and preventive services, supervision of programs of lower facility, training etc. Considering the enormity of work with the CHC, two major types of responsibilities are listed as follows:

1. Management function at facility level
2. Public Health function at filed level

1. Management at Facility Level

Management at the facility level is again divided into two categories which may be listed as follows.

a) Administrative function
b) Hospital Service Management function

a) Administrative Function

Human Resource. At every level of work under the CHC there is a considerable HR management challenge. It includes recruitment of ASHA, positioning of staffs, monitoring their work and administration of HR.

Finance. At the CHC level there are large amount of funds being transferred for various programmes under the process of decentralization under the NRHM. Also there are state funds which go down to the PHC, SC level and funds of SKS are also routed through CHC. It is important to ensure that funds are timely transferred, their proper planning and utilization is done and proper maintenance of accounts is being done.

Reporting: CHCs also generate and transmit large number of reports whose accuracy and timely transmission has to be managed. With increasing demand of computerization and HMIS and DHIS being fed at facility level the IT function of CHC is an important management function.

b) Hospital Management

CHC works as a small hospital and provide good clinical services which include OPD/IPD, 24X7 delivery services, diagnostic services etc. It also works as a secondary care for the referral of the lower level facilities. Hospital Management involves management of large number of support services which are essential for clinical services to work efficiently. They include management of laundry, sterilization services, housekeeping, cleaning and sanitation, security, Medical record management, Medical Gas Management, pharmacy management, blood bank services, OPD/IPD registration and queue management, ambulance management, bed management, repair and upkeep of machinery and equipments etc.

2. Public Health Service in the Field

CHC is an important level from which all public health initiatives have to be carried out. CHC acts as a hub for the various national and vertical programmes. The important National programmes which need special attention include the immunization, RCH, family planning, school health (RBSK), adolescent health programme, RNTCP, national vector borne disease control programme, non-communicable disease programme, AIDS control programme etc. CHCs have to do disease surveillance to prevent any outbreaks and take corrective actions in case of outbreaks. It also does the planning, implementation and monitoring of the programme to be rolled out for public health services.

Study on CHC

A study named, "Availability of Human Resource and Health Services and their utilization at CHCs of Haryana", conducted by Haryana State Health Resource Centre (HSHRC). This study aims to classify CHCs according to the population served, grade all CHCs as per its availability of services, human resource and service utilization and recommend strategies for policy change/modification and proper management of CHCs and propose to function as FRUs. The study was conducted in the state of Haryana and covered all CHCs in 21 districts. It is a kind of complete enumeration and quantitative study.

Study Outcomes

The study revealed that

- ☆ Only 13 per cent of the CHCs are covering population as per the prescribed norm (between the range of 80,000-1,20,000). In the other hand 20 per cent and 18 per cent of the CHCs served less than 80,000 and more than 2 lakhs population respectively.
- ☆ The availability of key health services and their utilization was found significantly low at CHCs.

☆ 59 per cent of the Health Service provider's positions are vacant in the CHCs in the state (specialists, medical and paramedical staffs and others).

☆ As per the grading of the CHCs done in terms of availability of the health services, human resource and utilization of services, it was found that no CHCs was in category A.

The underutilization of the CHCs may be because of:

☆ Lack human resource

☆ Lack of availability of health services at CHCs

☆ Less population coverage by some CHCs than norm (less than 80,000) may result in underutilization of the CHCs.

☆ Population coverage more than norm (more than 1,20,000) by some CHCs may result in more number of patients in the facility by which quality of the health services suffers.

Subsequently poor quality of the health services leads to underutilization of the CHC. As CHCs is also engaged in public health services like promotive and preventive services, supervision etc. other than the clinical services. It is very difficult to provide quality of care with adequate utilization, especially with scarce human resource. On the other hand it is also difficult to implement and monitor public health services in the field. Hence there is a need to rationalize the human resource of the CHCs according to population it covers and formulate a minimum of benchmark of services at CHCs. It can be stated here that the health services affected at both the places i.e clinical and public health services front.

Study Recommendations

Recommendations for the policy modification of the study are as follows:

1. Rationalization of CHCs according to the population coverage/ distribution.

2. Minimum benchmark of the health services for CHC should be fixed as per the categorization/rationalization of CHC.

3. Clinical and public health services should be segregated.

4. Rationalization of Human Resource as per the new categorization of CHCs and Health service.

5. CHC's role should be defined as per the clinical services, promotive and preventive services, supervision of programmes and trainings etc.

6. There is an urgent need of Public Health Cadre to manage the public health functions of the CHC.

Rationalization of CHCs

As per IPHS and GoI norm CHC should have a population coverage between 80,000 to 1,20,000. The study "Availability of Human Resource and Health Services and their utilization at CHCs of Haryana" revealed that there is large disparity in population distribution by the CHCs. There is need to ensure that population norms are followed and man power rationalized according to the population it covers (as the population coverage increases the load on CHCs increases). The following table shows the current status of existing CHCs, the study outcome of the type CHCs according to population coverage and proposed rationalization.

Table 6.1: Status of CHCs in Haryana

Existing Nomenclature of CHCs in Haryana	Categorization of CHCs as per the Existing Population Coverage (From the study)	Proposed CHCs as per the New Categorization (Proposed)
A	B	C
1. 30 bedded CHCs with PHCs in its jurisdiction 2. 50 bedded CHC cum hospital with PHCs in its jurisdiction 3. 50 bedded functional CHCs with no PHCs in its jurisdiction 4. 6 bedded Block PHC with PHCs in its jurisdiction	1. CHCs covering less than 50,000 population (14%) 2. CHCs covering population 50,000 to 80,000 (6%) 3. CHCs covering population 80,000 to 1,20,000 (13%) 4. CHCs covering population 1,20,000 to 2,00,000 (46%) 5. CHCs covering population 2,00,000 to 3,00,000 (13%) 6. CHCs covering more than 3,00,000 population (5%)	1. Mini CHC (Type 1)-10 beds 2. CHC (Type 2)- 30 beds 3. Rural Hospital (Type 3)-50 beds
Total no. of CHCs = 110	Total no. of CHCs = 110	Total no. of CHCs = 133

As mentioned in the table for the rationalization and bringing uniformity for criteria for establishment of CHCs in Haryana, It is proposed that the CHCs should be formed as per the population it covers. It should also be taken care of any city/ town comes in CHC area and any other health infrastructure in the area. There is a need to physically and functionally segregate the clinical and public health functions at CHC level. There are three type of CHCs are proposed with the criteria explained below.

Type 1 CHC-Mini CHC with 10 Beds

CHCs covering less than 50,000 population and 50,000 to 80,000 (A and B of above): (i)These CHCs will be called Mini CHCs which will have staff position more than PHC and less than CHCs.

1. It may have 10 beds. The population coverage by the CHC and utilization can be taken into consideration to decide upon the number of beds.
2. The public health part of the CHCs will be taken care by the Public Health Cadre at block or CHC level.
3. The details of health service availability and required Human resource for these CHCs are discussed in the annexure. The minimum required medical equipment list has also been prepared and can be seen at annexure.

Type 2 CHC-CHC with 30 Beds

1. CHCs covering population between 80,000 to 1,20,000 and 1,20,000 to 1,50,000 (C and D) to be left as it is. It will remain as the CHCs with 30 bed and supervision of the PHCs.

2. The CHCs covering population 2,00,000 to 3,00,000 and more than 3,00,000 population (E and F) will be split. The 2,00,000 to 3,00,000 covering population will be split to 2 to 3 CHCs, each covering population between 80,000 to 1,20,000. The CHCs covering more than 3,00,000 population should be split into 3 CHCs or more to make all the CHCs between 80,000 to 1,20,000 population coverage.

3. The public health part of the CHCs will be taken care by the Public Health Cadre at block or CHC level.

4. The availability of the services and required man power for Type 2 CHCs has been revised. The details are in the annexure. The minimum required medical equipment list has also been prepared and can be seen at annexure 4.

Type 3 CHCs-Rural Hospital with 50 Beds

1. The CHCs covering population between 1,50,000 to 2,00,000 (D) will be made 50 bedded hospital.

2. All these CHCs will be FRUs and provide the secondary level care including C-section services to pregnant women.

3. These CHCs will have slightly more Human Resource than the 30 bedded CHCs.

4. The public health part of the CHCs will be taken care by the Public Health Cadre at block or CHC level.

5. The detail of service provision and required human resource are in annexure. The minimum required medical equipment list has also been prepared and can be seen at annexure.

After rationalization of the CHCs, in Haryana there will be only three types of CHCs, *i.e.*

1. Mini CHC - 10 bedded
2. CHC - 30 bedded
3. Rural Hospital - 50 bedded

Additionally, the following points may also be considered for the proposed reformation of existing CHCs in Haryana to Mini CHC, CHC and Rural Hospital.

☆ If the currently has more than 2 lakh population coverage and it has town with more than 20,000 population the split of CHCs should be done where one rural hospital will also be created with 1,50, 000. The other CHCs may be a Mini CHC or CHC. The rural hospital (50 bedded) will be at the town where the population is more than 20,000 population (town alone).

☆ If the current population coverage of the CHC is as per the norm and there is town more than 20,000 population other than the CHC location, alternative arrangement like dispensary can be proposed there.

☆ If the current population of any CHC is between 1,50,000 and 2,00,000 and a town having more than 20,000 population comes in the area, it has be upgraded to rural hospital (50 bedded)

☆ For the towns or block head quarter town it has to be checked if there is any current Subdivisional Hospital or General Hospital is there or not. If yes then it may be left out to establish any Rural Hospital

The total number of CHCs (numbers are approximate, it may change with the consideration at state/district level keeping in mind the other criteria of establishment of CHCs of Type 1, 2 and 3) to be formed as per the new categorization are below.

Table 6.2: No. of CHCs as per New Categorization and their Rationalization

Sl.No.	Particulars Existing no of CHCs=110 (all type)	Type of CHCs		
		Mini CHC	CHC	Rural Hospital
1.	No of CHCs after rationalization (total=133)	27	84	22
2.	CHC+CHC *(by splitting CHCs more than 2 lakhs population)		26 (out of existing 10, 16 additional)	
3.	Rural Hospital + Mini CHC **(by splitting CHCs more than 2 lakhs population)	10 (10 additional)		10 (10 additional)
4.	Rural Hospital + Rural Hospital **(by splitting CHCs more than 2 lakhs population)			2 (out of existing 1, 1 additional)

*   A CHC which is currently population between 2-3 lakhs split into 2 to 3 CHCs with 0.8to 1.5 lakh population coverage.

**   A CHC which is currently population more than 2 lakhs and town with population 20,00 in it can be split into a Rural hospital and a mini CHCs.

*** A CHC which is currently covering population more than 3 lakhs can be split into 2 Rural Hospital as there are two towns having more than 20,000 population each.

The detail of the individual CHCs can be seen at Annexure

The result of the exercise will yield in evenly distributed CHCs across the state. It will help in rationalization of Human resource and helpful in monitoring of performance and utilization of the CHCs. It will also define certain level of defined benchmark of services to be provided and availed at certain level of CHCs.

Benchmark of Services for CHCs

As per the population coverage, the CHCs have been rationalized as three type of categorization. The minimum benchmark of service availability at all level has been defined and assured.The human resource required is being shown in the annexure.

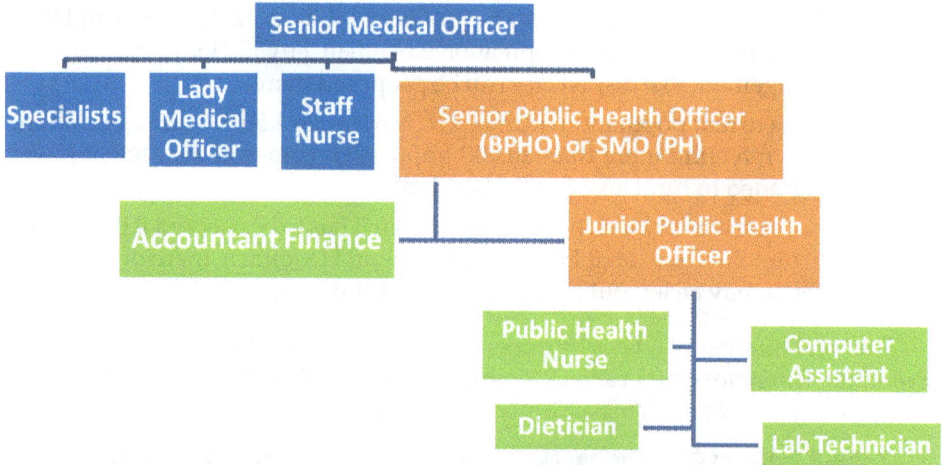

Figure 6.1: Proposed Flow Chart of the Public Health Cadre for the CHC.

The current staffs available will be utilized maximally to build the public health cadre at CHC level who will take care of all public health activities at block level. The Public Health personnel proposed at CHC level are:

☆ Block Public Health Officer (BPHO) or SMO (PH) – 1 -Additional/new post

☆ Junior Public Health Officer (JPHO) – 1 -Additional/new post

☆ Accountant Finance – will report directly to BPHO

☆ Dietician

☆ Lab Technician

☆ Public Health Nurse

☆ Computer Assistant

The CHC will take care of the clinical services where as the Public Health Cadre will take care of preventive care, promotive care, supervision of programs, training etc.

Chapter 7

Evaluation in India:
Retrospect and Prospects

K.N. Pathak

Joint Advisor,
NITI Aayog, New Delhi
E–mail: knp.pathak@gmail.com

Evaluation and Monitoring are the two fundamental aspects of planned development. To ensure that the development is taking place in desired direction and at the requisite pace, it is necessary that all the schemes are monitored methodically based on clearly indicated parameters.

Usually, most of the governments have systems to measure spending, processes and outputs. Not many use this information for assessing results produced by governments- its output, outcomes and impacts for each programme. This mainly happens because of an element of fear. While positive evaluation findings that reveal good performance are always welcome, adverse findings can pose significant political and reputational risks.

Impact evaluations are required to understand the effectiveness in addressing social problems of, both the government and privately operated programmes. Based on the past few decades of experience, it is felt that evaluation constitutes the core of development planning. Diagnostic study of any government scheme/programme helps in providing mid-course correction for betterment of the scheme and strengthens the hands of the government. Evaluation results can help decision makers focus resources on programmes that are relatively effective and, importantly, can provide evidence to improve the performance of key programmes that are not performing as well as might be hoped. An important prerequisite for evaluation is that it should be conducted with reference to a baseline data. But unfortunately, base line data is missing in case of a large number of the schemes/Programmes awaiting evaluation.

An effective and successful Monitoring and evaluation system can be ensured if good quality performance information and evaluation findings are generated and used at one or more stages of the policy cycle for better outcome. Hence, it is required that both supply and demand of M and E information have desired attributes. Thus, demand side ie; users of M and E information must include Ministers, Senior officials, Policy Analysts, programme managers, legislature and civil society and extent of use of M and E information in the policy cycle. On the other hand, supply side should include attributes such as the quality and reliability of monitoring data, the number and coverage of performance indicators, the types of evaluations conducted, the issues addressed and the reliability and timeliness of the evaluation. Experience with M and E systems shows that powerful incentives are important on the demand side for achieving a high level of utilization of the information they provide. These incentives, which can yield the desired attributes on both demand and supply sides could be in the form of rewards, deterrents and sermons, (statements of supports)[4].

There are a number of case studies available regarding the countries that have succeeded in achieving high levels of utilization of M and E information. The notable among them are; Australia, Canada, chile and Mexico.[5] We could have a brief view of them.

Australia

Driven by the federal Department of Finance (DoF), the Australian M and E system which was in existence from 1987 to 1997, was considered to be one of the most successful in the world. All the ministries were asked to evaluate all programmes every 3-5 years. By 1994, 80 per cent of the new budget proposals relied on evaluation results. Two-thirds of saving options relied on evaluation findings. Cabinet budget proposals, line departments and Audit office also made use of evaluation findings. However, with changes in the Government in 1996 and opposition from Bureaucracy (political reputation risks), this M and E system was abolished in 1996.

Chile

The M and E system introduced in 2000 is centrally coordinated by the Ministry of Finance. *Strategic definitions* were introduced to provide information about each Organization's mission, strategic objectives and products (public goods and services provided), and its clients, users and/or beneficiaries. Under M and E in Chile, the main tools, (many of which were redesigned in 2000) include:

1. **Monitoring tools** comprising of Strategic definitions, Performance Indicators,Comprehensive management reports and Programmes for management improvement.

2. **Evaluation tools** comprising of Government programme evaluations, Impact evaluations, comprehensive spending evaluations and Evaluations of new programmes.

During the period 2000-10 the Chilean M and E reforms have contributed to developing a "measurement"- oriented culture across central government Ministries and agencies. It is recognized by public servants in Government, and various academic and international assessments. The Ministry of Finance commissions evaluation externally to academics and consultants with standardized ToR and methodology. Evaluation. Information is used in Budget analysis and performance assessment of each Ministry and agency. It is also used for performance target setting and impose management Improvements.

Canada

Performance measurement, monitoring and evaluation have long been a part of federal governance in Canada (since 1969).The system comprises an evaluation policy, standards and guidelines, which underwent 3 modifications in last 30 years. However, the following changes have been prompted by adoption of "letting managers manage" and introduction of RBM accountability and transparency in government:

1. M and E units in most departments under central leadership ensures RBM and accountability.
2. There are well defined rules setting policy, standards and guidelines by TBS with assistance from CEE, a unit with TBS.
3. OAG audits functioning of the M and E system and presents results to the Parliament.

Mexico

Political changes in Mexico in late 1990sgenerated an increased demand for transparency and accountability. New legislations/institutions came up to strengthen M and E. The OPPOTUNIDADES (conditional cash transfer programme) introduced in 1990s had a built- in evaluation component. This M and E became a role model in Mexican public administration. The development of M and E in Mexico can be divided in two stages. The first in the period 2000to 2006 and second the post 2007.The first phase was characterized by good intentions but had a limited vision on the institutional capacity building required for the new evaluation efforts. Welfare/social sector programmes needed external evaluation as per budget law. But a number of teething problems encountered as there was no in-house capacity to frame ToR, identify suitable institutions and supervise evaluation works. However, Ministry of Social Development (SEDESOL) embraced the M and E agenda later and introduced impact evaluation methodologies in many programmes and formalized use of Log frame and feedback from M and E. In the second phase ie, post-2006 a strategic partnership between MoF, CONEVAL (National Evaluation Council) and SEP(Ministry of Public Management) emerged to ensure evaluation quality and utilization. They issued guidelines, created

Evaluation units in line ministries along with a systematic feedback mechanism.

South Africa

The Department of Performance M and E(DPME) set up by Republic of South Africa in January,2010 has introduced a number of initiatives including a focus on

12 government priority outcomes; the assessment of the quality of management performance of national and provincial departments; a new system of monitoring front-line services; a national evaluation system and; and a municipal performance assessment tool. These tools have contributed to a major increase in the availability of evidence for policy and decision making. Rapid recent progress is due to strong support at the onset from South Africa's President learning from international experience, and strong teams in DPME and the national treasury.

In the Indian context, an evaluation policy, it is yet to be formulated and approved by Government of India. However, the increasing pressure on national governments to be fiscally responsible, adopt prudent macro management, improve public sector performance and promote efficient delivery of services focused the attention on Evaluation Capacity Development (ECD) particularly in the public sector[6]. Since social and gender equity have been two important components under various Development schemes/programmes implemented by Government, the Programme Evaluation Organization (PEO) of the Planning Commission, as the leading Evaluation agency of the Government, has been endeavouring to ensure that the evaluation conducted under its aegis, inter-alia, have a clear perspective of social and gender equity in its evaluation. However, keeping in view the changing global scenario and the role of private players in overall development of the country, framing of an evaluation policy seems to be highly desirable. As it is being observed, the responsibility of the government, with the growing population, is spreading far wide in the field of social sector such as health care, education, water supply and other welfare programmes. No doubt, in these sectors also, the private organizations have a considerable stake. Still, the investment made by the government in these sectors is far more than the private agencies. The target or the beneficiaries covered under the schemes/programmes in these sectors, both through the government as well as non-governmental agencies or private players are identical. May be, their objectives also could have similarities to a considerable extent, but the financial perspective is natural to differ. While governmental education institutions have welfare as fundamental objective, the private sector institutions generate good revenue lying at disposal of the management. Hence, the evaluation of programmes/schemes being implemented in these sectors through public expenditure need to have a proper perspective which could be ensured by formulating an evaluation policy endorsed by executive as well as legislature of the country.

Further, if a development scheme or a welfare scheme is to continue for a long time, it is necessary that it is evaluated at least once after every five years during the lifecycle of the scheme. It is noted that the total plan allocation for various developmental schemes funded by Govt. of India is Approx. Rs.575000.00 crore during the Annual Plan 2014-15. If we take an average step up of 10 per cent per annum in the plan allocation, it would amount to about Rs.2700,000 crore for the full Five Year Plan. Thus, for a magnitude of this size of budget for development schemes of the country, we need to have a sound evaluation and monitoring system equipped with strong institutional mechanism. The resolution regarding National Institution for Transforming India (NITI) Aayog announced by the Government with the advent of the New Year 2015 also expresses the commitment *"To actively monitor and evaluate the implementation of programmes and initiatives, including identification of the needed resources so as to strengthen the probability of success and scope of delivery".*

Presently, we have about 1134 schemes in the Central Sector and 66 schemes under the Centrally Sponsored category (with a number of sub-schemes under one umbrella scheme in many cases). Among these schemes, there are more than 100 schemes costing Rs.300 crore each or more. Out of these, 17 schemes are flagship schemes which have nationwide coverage and generally they aim at benefitting the common people of the country. Merely 25 per cent to 30 per cent of these schemes are evaluated with a large scale coverage. In India, there is a progressively rising demand for unbiased and authentic information on the impact of public programmes. Such demand is being generated both from within the government and from outside. The implementing Ministries and Central and State Departments also have been laying emphasis on evaluation results as resource allocation and budgetary processes of development programmes are getting increasingly linked to performance and outcome[7]. However, as the existing evaluation agency of the Government; the PEO of the Planning Commission had a limited structure, the Ministries/Departments and States got various schemes evaluated through the different agencies. Since the agencies conducting the evaluation get the fund for the evaluation from the implementing Ministry or the Department itself, they have too many constraints. What is generally seen that the evaluation report submitted by most of the external agencies (which is called a *'third party evaluation'*) often becomes a honey coating exercise particularly, if it is funded by concerned implementing Ministry or Department. An evaluation assigned on contractual basis may not guarantee objectivity basically because of the dependence of the evaluation agency on the sponsoring Ministry/Department for funding as well as logistic support. If the evaluation report is really objective and dares to bare the truth, the agency can never expect to get the evaluation assignment from any Ministry or Department in future. Moreover, as the fund for evaluation is provided by the nodal Ministry/Department, major drawbacks/deficiencies in implementation mechanism of all such schemes are either ignored or covered up in evaluation.

The monitoring and evaluation of development interventions in India came more or less along with the concept of planning. The 1st Five Year Plan document stated that "Planning is essentially an attempt to coordinate means and aims. A planned economy has in view a somewhat wide time horizon to which day to day decisions have to be related.... Practical policy cannot operate in terms of mere set of doctrines; it must satisfy some pragmatic tests."[8] The need of strengthening the planning and implementation was reflected from very First Five Year Plan itself with the government making a provision of Rs.50lakhs in the First Plan for research and investigations into selected economic, social and administrative problems of national development in cooperation with universities and research institutes.

In the last six decades of its existence, the Programme Evaluation Organization of the Planning Commission has conducted 220 studies ranging from the issue of community development to health care, education, employment, irrigation, food and civil supplies, local self-governments etc. There is no other institutions/organization in the country which either has such a long history of evaluation or has any agency conducted evaluation on such a wide range of issues/disciplines. However, this organization is also understaffed now. Once upon a time, it had 32 field offices in the country which are presently 15 in number. If the evaluation has to be conducted in a time bound, objective and scientific manner, this organization needs to be expanded and provided requisite wherewithal to conduct evaluation spread over vast disciplines and covering the entire country. Such an organization may have the scope of co-opting experts from the open market either on the pay roll or on consultancy/contractual basis. A vibrant organization with mandatary strength and vast expertise can certainly be expected to undertake quality evaluation resulting in midcourse correction and modification in various governmental schemes/ programmes aiming at peoples' welfare and development of society in general. To adhere to evaluation ethics, the evaluation organization also needs to be given legal support and mandatory strength so that every institution running through public money is legally bound to share the information solicited by the Evaluation Organization.

While formulating the evaluation policy, the following related aspects also need to be kept in view:

1. Administrative and Structural Issues

The evaluation agency should be independent (insulated from political/ bureaucratic interference) and accountable to Parliament. The Head as well as all the Members of such evaluation body should have profundity in the field of evaluation and monitoring. The mode of selection/appointment of the Head of the Independent Evaluation Body and the entire personnel structure of the evaluation body should be same as that of CAG. The funding of Independent Evaluation Body should be identical to CAG or so.

Moreover, by having an independent evaluation agency, a platform could be created for career progression of evaluation experts. It is not necessary to keep this agency as a watertight compartment. Rather, there should be a provision to have lateral entry of evaluation experts into evaluation agency at different stages.

The laying of every evaluation report by the evaluation agency on the Table of Parliament may be made mandatory like the CAG report. This would enable the Members of the Parliament to have a wider and deeper insight into different schemes being implemented by the government.

The training infrastructure of evaluation is also very necessary. The need and justification for constituting a working group by Government of India for strengthening the training infrastructure on evaluation under the Chairmanship of late S.S. Puri in 1979 is all the more necessary even in the present time. The Core Group constituted by CAG, in its Report in 2003 also had suggested that a pool of evaluators trained in new techniques need to be created in order to establish a credible supply base.

i) Financial

Financial provision for the evaluation agency enjoined with the financial autonomy may be expected to give a smooth functional platform for the organization.

Though, every kind of expenditure incurred under the institution would be as per the General Financial Rules, it would be free from routine bureaucratic hurdles which delay the field work or data collection analysis for want of resource provisions.

There is also a thinking that if a corpus fund should be created through a new initiative, to provide support funds to research institutions, universities, NGOs, Governments, development agencies etc., to conduct impact evaluations. The institution controlling the corpus could offer grant to assist in the design of impact evaluation or, depending on decision by stake holders, it could have resources for financing impact evaluations with a grant to be awarded on an open competitive basis. This idea was floated by Center for Global Development a few years ago and there is scope for serious discussion on this issue.

ii) Collaborative

Reviews of impact evaluation have come out with the view that there are not may impact studies being conducted with the requisite reliability and validity to provide the evidence base needed by policy makers in developing countries. Therefore, the view has been raised in many quarters to take up collective action by multilateral agencies, bilateral agencies, private foundations, NGOs and research centers. Such an approach, it is believed, will bring sufficient investment in impact studies and the findings will be widely disseminated and data will be made public and also the studies would address questions enduring importance and relevance to policy makers. It is also believed that collaborative approach in evaluation would address the questions of enduring importance, provide models of good practice for emulation; and would also promote methodological innovation and high evaluation standards.

The Evaluation Gap Working Group constitute by the Center for Global Development in 2008 suggested that a pioneering group of governments, international agencies and private foundations create a new entity that would focus on impact evaluations. This entity, according to CGD, would lead the definition of share questions that are of interest across countries and agencies; provide funds for the design of evaluation at the all-important moment when programmes are being designed; develop quality standards for evaluation; disseminate the results of good evaluation and undertake other core functions.

Many are of the view that evaluation should not be made the solitary prerogative of the government. We may agree that social audit and participation of private organizations and civil society in evaluation is very necessary. But it cannot be denied that any evaluation agency cannot afford to effectively pinpoint the fallacies underlying a particular scheme or programme of the sponsoring Ministry or Department. There is always the fear that *"those who pay the piper call the tune"*. Thus, the lack of financial autonomy and mandatory strength may not ensure full strength for an evaluation agency. Our emphasis on this issue should not be mistaken as the plea for keeping the academia or civil society or NGO out of the ambit of evaluation. What is actually desirable is that for conducting methodically sound and ethically appropriate objective evaluation; the evaluating agency should not be dependent on any governmental department or client agency for providing funds. One cannot disagree from the view that for effective evaluation, the involvement of outside experts consisting of academia, NGOs and civil society at different stages of evaluation is very necessary. There is certainly the need for "methodological pluralism" to address evaluation questions in different contexts. However, there is a long way to go to achieve the same in case of India.

It is high time to introduce Evaluation and Monitoring as an academic discipline. This will equip the young researchers, executives and civil society workers with the complete methodology, tools and techniques as well as ethical components of evaluation and monitoring. Since we do not have any course curriculum on this account, except for a diploma course being run by IAMR of the Planning Commission (now renamed as NITI Aayog), it qualifies to be given due consideration.

iii) Ethical

Adherence to evaluation ethics is a basic requirement for an objective evaluation. Unless we have a truly independent evaluation agency (which does not have to depend on the client agency for funding), the evaluation exercise would be superficial. Hence, it is desirable that India has a strong and independent evaluation agency with constitutional safeguard like CAG. Such an agency should be accountable to the Parliament. Providing such a wherewithal and legal teeth to Apex level National evaluation agency can certainly ensure effective implementation of a large number of development schemes being implemented by Govt. of India.

However, Mere ethics will not do. What is essential is that an efficient independent and financially sound evaluation agency committed to adhere to the ethical principles of evaluation is given all support. It is highly desirable that to make India a global leader, all our development schemes, particularly, the flagship ones are implemented as per the envisaged objectives and the desired outcomes. This could be ensured by conducting methodologically sound evaluation which needs a structure and the functions as elaborated above.

References

1. MacKay, Cited in Lopez-Acevedo (2012), p. 6.

2. Lopez-Acevedo (2012) Gladyz, Philipp Krause and Keith Mackay (Ed), (2012): Building Better Policies: The Nuts and Bolts of Monitoring and Evaluation Systems, World Bank, Washington D.C. p.8

3. Comptroller and Auditor General of India: Development Evaluation in India – Contemporary issues and approaches; Report submitted by the Core Group February, 2003.

4. CAG oppcit. p.2

5. Planning Commission, First Five-Year Plan.,p.7-8.

Other References

1. Closing the Evaluation Gap: *Frequently Asked Questions, Center for Global Development, Electronically retrieved from website www.cgdev.org/pge/frequently-asked-questions-2 on 30.12.2014.*

2. Comptroller and Auditor General of India: *Development Evaluation in India – Contemporary issues and approaches; Report submitted by the Core Group February, 2003.*

3. Development Evaluation Society of India (DESI): Home Page – electronically retrieved from website www.desiindia.org/evaluation-system.html on 30.12.2014.

4. Lopez- Acevedo, Gladyz, Philipp Krause and Keith Mackay (Ed), (2012): Building Better Policies: The Nuts and Bolts of Monitoring and Evaluation Systems, World Bank, Washington D.C.

5. Mackay, K (1997): Evaluation Capacity Development, World Bank, ECD, Working Paper No. 6.

6. NIC, Government of India: *NITI Aayog: Objectives and Composition,* electronically retrieved from webmail.nic.in/frame.html? and security=false and lang=en on 1.1.2015

7. Planning Commission, Government of India:

(i)   India's Planning Experience: A Statistical Profile, 2001

(ii) Development Evaluation in PEO and its Impact, 2004 and 2006

(iii) Proceedings of the Conference of Central and State Evaluation Organizations held in New Delhi on 28th July, 1999 and their Status Papers; August, 2000.

Chapter 8

Programme Evaluation in India and the World: Need for a National Evaluation Policy

B.K. Pandey[1] and Abhishek Mukherjee[2]

[1]Advisor, [2]Economic Officer
NITI Aayog,
Room No. 312, Yojana Bhawan, Sansad Marg,
New Delhi – 110 001
E-mail: [1]bk.pandey@mail.gov.in
[2]abhishek.mukherjee@mail.gov.in

ABSTRACT

The evaluation of public programmes and performance is immensely important as an engine of development. Evaluation of policies shows the path and highlights the focal points where more attention is to be provided. Proper evaluation and monitoring is essential for judging the impact of any policy or programme. The paper deals with the importance of evaluation, the mechanism of evaluation as it has evolved in India, practices in western countries, more importantly countries of Africa and South Asia and the need for a national policy on evaluation in India.

Keywords: *Evaluation, Programme evaluation, Performance evaluation, Social audit, Outcome budgeting, PEO, PMES, TBS, SLEvA.*

Introduction

Over the past years, there has been a growing trend towards the better use of evaluation to understand and improve practice. United Nations General Assembly latest resolution in Sept 2014: Empowering Countries through Evaluation: Evaluation as a country-level tool for the new development agenda emphasizes the importance of national-level evaluation capacity building, use of evaluation for evidence-based decision-making and to strengthen their national evaluation policies and systems.

In recent years, India has taken a number of initiatives to improve the monitoring and evaluation environment such as the Outcome, Gender and Green Budgeting by Ministry of Finance, and the Performance Management and Evaluation System (by the Cabinet Sectt). Although, there is high level of evaluation capacity and professionals in the country, yet there is no national policy on evaluation nor any central legislation to that effect as in many developed countries.

This paper discusses the India's approach to evaluation of programs and system of evaluation. It is organized into IV sections. After introduction, section I underlines the evaluation as a tool of better governance and about the need for strengthening the evaluation system in the country, Section-II gives an outline of the evaluation system in India including States. Section III provides an overview of evaluation system in some western countries, developing countries of Africa and South Asia in particular. Section-IV concludes by suggesting a need for a national policy on evaluation.

Evaluation as a Tool for Better Governance

It has been rightly said that "Program evaluation is the systematic collection of information about the activities, characteristics, and outcomes of programs to make judgments about the programme,improve programme effectiveness, and/or inform decisions about future programming." (Patton, 1997). Evaluation uses systematic data collection and analysis to address questions about how well government programs and policies are working, whether they are achieving their objectives, and why they are, or are not, effective. It produces evidence that can be used to compare alternative programs, guide program development and decision making. Evaluation can provide vital information to executive, legislative and judicial wings about the outcomes, their effectiveness, efficiency, and worth and thus provides a useful and important tool to bring credible, well-grounded information to bear on a broad range of government decisions allowing for better governance.

Programme Evaluation and Performance Evaluation

There is a slight distinction between programme evaluation and performance management, although they support each other. In practice, the term program has been used to refer to a Government policy, activity, project, initiative, law, tax provision, function, or set thereof. (Thousand Oaks, CA: SAGE Publications, 2005, p. 139).The term performance measurement on the other hand is usually considered different from program evaluation. Performance measurement refers to ongoing and periodic monitoring and reporting of program operations or accomplishments (*e.g.*, progress toward quantitative goals), and sometimes also statistical information related to it.

Lack of a Coordinated Evaluation Policy

Over the years, the evaluation science has developed an extensive array of analytic approaches and methods that can be applied and adapted to a wide variety of programs, depending on the program's characteristics and implementation stage, the way the results will be used, and the kinds of decisions that will be made. One has to find out methods that are appropriate to assess a given type of program or policy, whether evaluating agencies have sufficient capacity to evaluate their programs and have sufficient independence to credibly evaluate the programs and policies. But for the most part, these evaluations have been sporadic, applied inconsistently, and supported inadequately. Many a times, the units formed to conduct evaluations are under-resourced and not given due importance. Training and capacity building for evaluation have been inadequate across agencies especially at the State level and, in many cases, insufficient to achieve the needed evaluation capacity and sustain it over time. Despite an increased understanding of the need for - and the use of - evaluation, however, a national policy on evaluation has been lacking. These are some of the questions which needs to be debated and addressed. In this context, this paper seeks to analyze the need for a national evaluation policy and legislation, quality standards, capacity building, etc to give evaluation its rightful role in India development endeavor.

Evaluation System in India

Since the inception of Planning Commission in 1950s, evaluation has been considered to be an essential feature in the domain of public governance. Planning Commission had an evaluation unit as PEO to undertake evaluation work to assess the process and impact of development program, identifying the areas of success and failures at different stages of administration and execution. Planning Commission also promoted State Evaluation Organisations (SEOs) in the first two decades of planning to strengthen the planning machinery at the State level (U.P. was the first state to have an SEO in 1953-54).It also Constituted a Working Group (1964; Chairman: V.K.R.V. Rao) and a Task Force (Chairman: B.S. Minhas, 1972) for strengthening evaluation machinery of the states and another Working Group (Chairman: Dr. S. R. Sen, 1967) and a Committee (Chairman: S. S. Puri Committee, 1979) for strengthening the training infrastructure on evaluation;. Constituted a Committee for Review and Strengthening of Central and State Evaluation Organisations (the Dubhashi Committee, 1980). Figure 8.1 shows the system of evaluation in the Planning Commission:

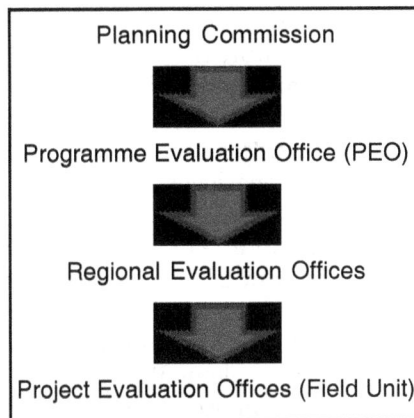

Figure 8.1: System of Evaluation in Planning Commission.

The Four Major Steps of Working of PEO is shown in Figure 8.2.

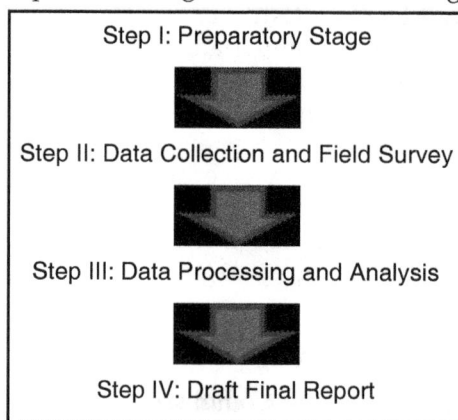

Step I: Preparatory Stage

Step II: Data Collection and Field Survey

Step III: Data Processing and Analysis

Step IV: Draft Final Report

Figure 8.2: Working of PEO.

An Independent Evaluation Office (IEO) was set up in 2014 to assess independently the impact of Government's flagship programmes. IEO has since been dismantled, however, the formation of "National Institution for Transforming India" (NITI Aayog) there, will give greater emphasis on monitoring and evaluation of various Governmental schemes as mentioned in the Cabinet resolution issued on 01.01.2015. One of the main objective of 'NITI Aayog' would be "to actively monitor and evaluate the implementation of programmes and initiatives, including the identification of the needed resources so as to strengthen the probability of success and scope of delivery".

Evaluation Units and Practices in Central Ministries

Most of the Central ministries have their internal evaluation and monitoring divisions to carry out the evaluation and there is also a provision of funding for this purpose in almost every government schemes. System of Cost benefit and cost effectiveness analysis, public expenditure tracking surveys, participatory methods, third party evaluation, rapid appraisal methods and impact evaluation have been used in India in varying degree. Social Audit, a participatory method of evaluation and has been widely used in assessing Mahatma Gandhi National Rural Employment Guarantee Act (MGNREGA)- the national level scheme for guaranteeing employment in rural India. Such involvement in evaluating public schemes has been successful in exposing corruption. Indian state of Andhra Pradesh has particularly been successful in proper implementation of Social Audit. It has set up an independent body Society for Social Audit, Accountability and Transparency. Annual Common Review Mission has been one of the important monitoring mechanisms under NRHM. The method of Outcome budgeting has also been implemented by all the ministries of Central Government as per the instruction of Finance Ministry in 2005-06. However, outcome budgeting has not been very successful in India, primarily because of lack of knowledge and skill of the ministries concerned to define their intended outcomes and ministries need to be further guided and nudged in this regard.

Performance Management Evaluation System (PMES)

Concurring with the recommendation of the India's second Administrative Reforms Commission (ARC) 2005, the President in her address to the Parliament in 2009 observed that the government should initiate steps towards establishing mechanism for performance monitoring and evaluation on a regular basis. On September 11, 2009 Prime Minister announced the introduction of PMES PMES has been introduced to evaluate and monitor the functioning of Government departments on the basis of work accomplished to the annual target that is being set. The evaluating organization judges the achievement on the basis of a scale ranging from excellent to poor. Monitoring involves keeping a tab on the periodical progress towards the annual target that is being set.

India has made a good beginning by creating the Performance Monitoring and Evaluation System (PMES) currently being implemented which covers 80 departments and 800 responsibility centres. In addition, 15 states, have adopted this system. However, this is not to be equated with programme evaluation. The Figure 8.3 gives the structure of PMES:

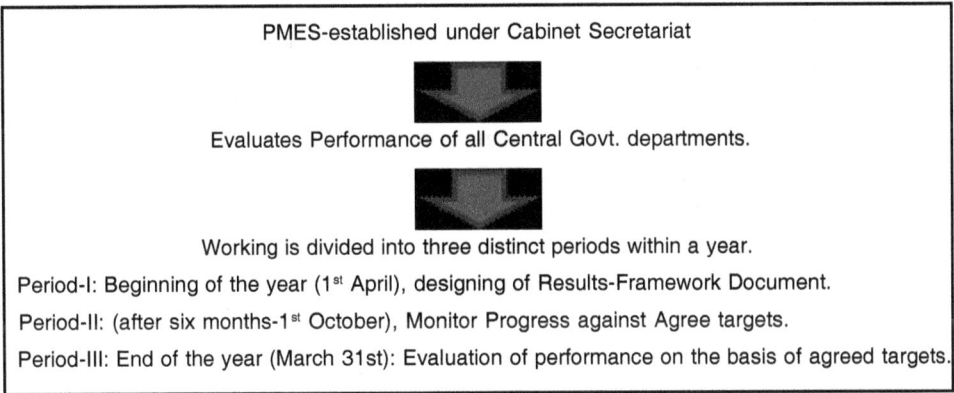

PMES-established under Cabinet Secretariat

Evaluates Performance of all Central Govt. departments.

Working is divided into three distinct periods within a year.

Period-I: Beginning of the year (1st April), designing of Results-Framework Document.

Period-II: (after six months-1st October), Monitor Progress against Agree targets.

Period-III: End of the year (March 31st): Evaluation of performance on the basis of agreed targets.

Figure 8.3: Structure of PMES.

Evaluation system in the States

Karnataka is the first state in the country to take a lead by establishing Karnataka Evaluation Authority (KEA) in 2011. The policy requires all the government departments, corporations, boards, local bodies and other publicly funded entities to get their plan as well as non-plan programs and schemes evaluated once in five years. Continuation of the plan programs and schemes is contingent on the evaluation outcome. Key objective of the Authority is to promote transparent, effective, efficient evaluations of public policies, programs and services. It promotes two types of Evaluations:

a. Internal evaluations commissioned by respective departments/entities, and

b. External evaluations commissioned by KEA with its own resources.

Evaluation Policy/Acts/Practices in Western Countries

In the United States of America, the last two decades have noted an increased interest in outcomes-based performance monitoring and evaluation (M and E). Implementation of the Government Performance and Results Act of 1993 (GPRA) under President Clinton, the President's Management Agenda and Program Assessment Rating Tool (PART) under President Bush and most recently, the High Priority Performance Goals (HPPG) Initiative and the Program Evaluation Initiative under the President Obama have brought greater emphasis on transparency and results. The United State's governance system is highly decentralized, there is no single unified and comprehensive M and E framework, and, as a result, there are a wide range of entities monitoring and evaluation programs across major departments and federal agencies and institutions. Apart from government departments such as Office of Management of Budget:, including the 15 cabinet departments, Congress also engages in M and E of government programs – both directly through its internal committee hearing process and indirectly through requirements that it imposes through legislation it enacts and through its own support agencies, principally the Government Accountability Office (GAO), the Congressional Budget Office (CBO), and the Congressional Research Service of the Library of Congress. A key feature of the U.S. M and E environment is the vigorous role played by non-governmental organizations both in overseeing and assessing government policies and programs. The U.S. has introduced many techniques such as Randomized Controlled Trial (RCTs) and Programme Assessment Rating Tool (PART) to carry out its evaluation process and it has a Central Coordinating Unit.

In Canada, under the Treasury Board Secretariat (TBS), the Centre of Excellence for Evaluation (CEE) has been created since 2001.The CEE provides functional leadership, including advice and guidance in the conduct, use and advancement of evaluation practices across the federal government. In 2009, the evaluation policy of Canada was revised to ensure that the evaluation function can provide adequate programme effectiveness information to support the Central Expenditure Management System (EMS). In Australia, program evaluation became a crucial element in the budgeting process.The budgetary levers of the Department of Finance and Administration show that budget management can be a critical part of the overall policy and expenditure management system. The British system demonstrates that a strong commitment to performance management eventually requires attention to program effectiveness. The government has increasingly placed top priority on improving outcomes from the delivery of public services.

Evaluation in Developing Countries: Africa and South Asia

Africa

Developing country evaluation leaders have articulated the need for a new approach to evaluation and the role it plays in improving the wellbeing of humankind – in particular, the lives of the poor and vulnerable in developing countries. At the January 2012 gathering of the Africa Evaluation Association's biannual conference in Accra, Ghana, African evaluation leaders and policy makers highlighted five steps to be taken if they aspire to play a meaningful role in social transformation.

1. Broaden the inclusion of key stakeholders in evaluation.
2. Regard evaluative knowledge as a public good and share it widely.
3. Address evaluation asymmetries between developed and developing regions.
4. Broaden the objects of evaluation to learn more beyond the individual grant or project to a more strategic assessment of portfolios of investments, policy change, new financing mechanisms, and sector-wide approaches that tell us more about what works and what does not in different contexts.
5. Invest in the development and application of innovative new methods and tools for evaluation and monitoring that reflect multidisciplinary and systems approaches to problems and complexity

South Asia

The Community of Evaluators (CoE) is a platform that facilitates knowledge exchange between parties interested in evaluation in the South Asian region. In Sri Lanka there is a strong evaluation culture with civil society participating in evaluation through the Sri Lanka Evaluation Association (SLEvA). They work closely with the government of Sri Lanka in strengthening evaluation policy in the country. It also runs professional capacity building workshops and international conferences. Pakistan Evaluation Network (PEN) also has long years of experience working with evaluation professionals and policy makers. COE- Bangladesh, COE- Nepal, COE- Afghanistan and Development Evaluation Society of India (DESI) are the other country evaluation networks at country level. Teaching Evaluation in South Asia (TESA) is another regional initiative in South Asia to enhance professional development in the region. South Asia Parliamentarians Forum has also added contribution to the evaluation policy development efforts of South Asia. The goal of the Forum is to advance enabling environment for nationally owned, transparent, systematic and standard development evaluation process in line with National Evaluation Policy. However, the practise of evaluation is yet to be fully institutionalized in South Asia.

Need for a National Level Evaluation Policy

Many of the public funded programmes don't achieve the goals and objectives that these programme aspire to achieve. Evaluation practitioners blame lack of national policy for this lapse. Recently, a group of committed Parliamentarians from South Asian countries made efforts of establishing national evaluation policies in their respective countries and ensure transparency and accountability in public sector development projects. In India, every department has an internal evaluation and monitoring system but it does not have a central coordinating agency as Canadian system has. India has introduced a centralized (PMES) for performance management across different ministries which is laudable and complements the programme evaluation framework.

A strong evaluation culture can enable leaders and civil society to develop and support better policies, implement them more effectively, safeguard the lives of people, and promote well-being for all. None of the South Asian countries have a national evaluation policy in place, despite each country having a fairly satisfactory monitoring and evaluation mechanism in place in their respective public sectors.To make programme evaluation effective, a number of steps are required, such as assigning experienced, senior evaluation experts at high levels, sufficient and stable funding to support professional capacity building, well laid out evaluation policies across central ministries that can guide evaluation efforts and help ensure their quality. Further, we need to adopt quality standards to guide evaluation functions, create clearinghouses to share information about effective and ineffective program practices.

Mapping exercise to map out status of national evaluation policies/mechanisms/guidelines at country level with identifying success stories in selected countries may be taken up. NITI Aayog can initiate to collaborate with States like -Karnataka to share their experience in evaluation with other States. It is high time to frame a National Evaluation Policy incorporating evaluation standards, ensuring that evaluation standards are complied with, and institutional arrangements for facilitating follow-up action on reports, accessibility of evaluative information by the civil society, strengthening research, training and application of economic/statistics in evaluation. A central coordinating unit in the NITI Aayog should be created to provide leadership, human resource and training. A National level policy can be framed by keeping in view the best practices of other countries adapted to India's need.

References

1. Books

Deborah M. Fournier, "Evaluation," in Sandra Mathison, ed., Encyclopedia of Evaluation (Thousand Oaks, CA: SAGE Publications, 2005), p. 139.

Patton, M. (1997). Utilization-focused evaluation (3rd ed.).(Thousand Oaks, Sage Publication).

Rogers P., Petrosino A., Huebner T.A., Hansi T.A. December 2000:"Program Theory Evaluation: Practice, Promise and Problems".

Taylor M.,Purvdue D.,Wilson M.,Wilde P. 2005: " Evaluating Community Projects: A Practical Guide", Joseph Rowntree Foundation,York.

Wang V.C.X. August 2009: "Assessing and Evaluating Adult Learning in Career and Technical Education" Zhejiang University Press.

2. Journals

Nagel S. 2001 "Conceptual Theory and Policy Evaluation" Public Administration and Management: *An interactive journal* page 71-78.

3. Reports

Lahey R. November 2010 Evaluation Capacity Development:" The Canadian M and E System: Lessons learnt from 30 years of Development" ECD Working Paper Series No.23,The World Bank, Washington D.C.

Mehrotra S. October 2013 Evaluation Capacity Development:" The Government

Monitoring and Evaluation System in India: A work in progress" ECD Working Paper No.28, The World Bank, Washington D.C.

United Nations General Assembly Resolution, September 2014: "Empowering Countries through Evaluation: Evaluation as a country-level tool for the new development agenda".

4. Websites

Pal, S.P. 2011a. Evaluation of India's Evaluation System. http://www.desiindia.org.

Pal, S.P. 2011b. Development Evaluation in India: An Overview. http://www.desiindia.org.

Pal, S.P. 2012a. Policy Impact of Evaluation Studies – The Indian Experience.

http://www.desiindia.org.

Pal, S.P. 2012b. Evidence of Policy Influence of Evaluation Studies: The Indian Experience.

http://www.desiindia.org.

Karnataka Evaluation Authority, Government of Karnataka at www.karnataka.gov.in/KEAbangalore/Pages/home/aspx

Programme Evaluation Organization (PEO) at www.planningcommission.gov.in/rti/doc_rti/divi_details/peo.pdf

PMES 2014-15 at www.performance.gov.in

PARTICIPATORY APPROACHES TO EVALUATION

Chapter 9

Study of Administrative Structure of Hospital Management in 200 Bedded District Hospitals of Haryana

Ashish Gupta

*Haryana State Health Resource Centre,
Panchkula, Haryana
E-mail: hshrcpkl@gmail.com*

ABSTRACT

HSHRC had done a study on administrative structure of 6 hospitals from private and government sector and formulated an administrative structure for a 200 bedded district hospital. On the basis of this study, revised manpower norms were proposed and also some additional posts were proposed in order to strengthen administrative structure of district hospital. The hospital structure proposed by HSHRC has been approved by State government and Finance department. The objective of the study was to improve the quality of support services and outsourced services in the district hospitals in the state. The methodology of the study involved studying the outsourced services of 6 hospitals from private and government sector. A prospective study was carried out by visiting four private hospitals namely Fortis Mohali, Max Hospital Saket, Ivy Mohali and Medanta the Medicity, Gurgaon and two government hospitals that is AIIMS and PGI Chandigarh. Structured interviews were conducted with the respective facility heads about the organizational structure and administrative services in their facilities. The result of the study was that there are certain posts that did not exist earlier had been proposed and have been approved by Finance department, Government of Haryana, to provide better service delivery. These include:

Two Deputy Medical Superintendents (DMS-1 and DMS-2), One Clinical Psychologist, One Senior Pharmacist, One Senior Lab Technician, One Biomedical Engineer, One Quality Manager, Ten Data Entry Operators, One Dental Hygienist, Three ECG Technicians, One Plaster Technician.

The (Principal Medical Officer) PMO would be assisted by two Deputy Medical Superintendents (DMS-1 and DMS-2) and a Quality Manager in his/her administrative work.

DMS-1 would look after clinical services DMS-2 would look after non-clinical services and the Quality Manager would look after all the quality Improvement activities. Four supervisors each may be hired by Government on contract to assist DMS-1 and DMS-2. A retired JE/SDO would be hired to work under DMS-2 who would also be assisted by Bio-Medical Engineer, Chief Security Officer/Chief Fire Officer (CSO/CFO). Services like housekeeping, laundry, kitchen and security would be outsourced at district hospitals.

Introduction

Managing a modern hospital is like managing an industry with large number of varied functions and requirements. The advent of corporate hospitals in the country has raised the level of hospital care and management to new professional heights. However, in terms of workload the District Level Hospitals handle much more work as compared to the corporate hospitals. They also provide very good clinical services. But the Government hospitals are very weak in terms of support services, cleanliness, infection control practices, grievance handling, public relations etc. With the State Government taking major initiative on quality, the administrative load on the hospital management has increased many folds. There is an urgent need to evolve a professional management structure for the hospitals in order for them to meet this challenge.

A comparative study of structure of 6 hospitals from private and government sector was carried out in order to arrive at the list of services and structure required to manage these services for the district level hospitals.

Objective of the Study

To conduct a study of the administrative structure and outsourced services of the hospitals in private sector in order to improve the quality of support services and outsourced services in the state.

Methodology

A prospective study was carried out by visiting four private hospitals namely Fortis Mohali, Max Hospital Saket, Ivy Mohali and Medanta the Medicity, Gurgaon and two government hospitals that is AIIMS and PGI Chandigarh. Structured interviews were conducted with the respective facility heads about the organizational structure and administrative services in their facilities. Detailed study was conducted at Fortis, Max and Ivy hospitals; while at Medanta only the outsourced services were analyzed as per the information provided by CMO, Gurgaon.

Observations

Organizational structure of a hospital is a combination of a hierarchical and divisional structure, since there is a chain of command where some levels are under another level, but employees are organized in departments or divisions that have their own tasks. The topmost level of the administration services consists of the people who usually own and operate the hospital as a business enterprise. They are responsible for imposing policies and the budget according to the needs of the patients and employees. These people are referred as board of directors.

The general administrative services of the hospital are headed by a Chief Executive Officer (CEO) or Facility Director (FD) who has the day-to-day responsibility for managing the hospital business. He or she is the highest ranking administrative officer and oversees an array of administrative departments concerned with medical, non medical, nursing, fiscal, operations, public relations and personnel. The medical division being the major component is looked after by the Medical Superintendent (MS) who is in turn supported by a Deputy Medical Superintendent (DMS). The entire nursing unit is being sought by the Nursing Superintendent who directly reports to the MS. The other non clinical and operational activities are looked after by a Facility Manager or a Hospital Administrator. The Financial Controller directs fiscal activities of the hospital and the personnel are managed by head of the human resources department.

The major role of a Medical Superintendent (MS) is to recommend to the CEO or board of directors the appointment of physicians or doctors. Another function is to grant privileges and to provide oversight and peer-review of the quality of medical care provided in the hospital. The heads all the medical specialties such as medicine, pediatrics, surgery etc directly report to the MS. The MS is supported by a Deputy Medical Superintendent (DMS) who ensures coordination within medical and non medical services to deliver on service quality. The number of DMS may vary from one to two depending upon the bed strength of the hospital. The DMS is in-charge of the junior doctors, dietician, therapists and the other paramedical staff. Medical records department (MRD) and Central Sterile Supply Department (CSSD) are also assigned to the DMS.

Medical Records Department (MRD) is responsible to maintain the records of all inpatients and outpatients in the hospital. A MRD is managed by a Medical Records Officer (MRO) and Medical Records Technicians (MRT) who are trained in coding of diseases as per international classification *i.e.* ICD. Usually one or two class IV or GDA are also employed in the MRD to assist in the daily functioning and record keeping.

Central Sterile Supply Department (CSSD) is managed by two – four CSSD technicians depending upon the hospital bed strength. The main function of this department is receipt, cleaning, sterilization, packing, labeling and issue of materials, supplies and equipment, dressings and other specialized surgical items.

Other than the medical departments there are a variety of non-medical departments which also play a vital role in delivery of care. These include pharmacy, purchase, front office, laundry, kitchen, housekeeping, engineering, security etc. These departments are also managed by a DMS in some of the hospitals. A majority of these departments are outsourced in the private hospitals which primarily include kitchen, laundry, housekeeping and security. These non-medical departments and their daily functioning are managed by a Facility Manager or Hospital Administrator.

Each non medical department is headed by its respective HOD *e.g.* pharmacy is headed by chief pharmacist, engineering is headed by senior engineer who in turn report to the Facility Manager. The Facility Manager is also responsible to liaison with the outsourced agencies and manages their contracts. The outsourced departments are also overlooked by the hospital supervisors to monitor the outsourced staff and coordinate their activities.

Housekeeping, laundry, horticulture and pest control are usually managed by one Housekeeping manager who is supported by supervisors per shift. The outsourced staff in housekeeping includes GDA, Class IV, ward servants, cleaners, sweepers etc. The laundry is outsourced by three ways. First the laundry can be fully outsourced to an agency and can be functional in the hospital premises. Second the space and equipment both can be provided by the hospital and just the manpower can be outsourced. Third the laundry can be outsourced to an agency that cleans and washes the linen outside the hospital premises and only the linen store is maintained by the hospital.

Similarly kitchen is also supervised by the hospital level supervisor along with a dietician. Either the kitchen services are fully outsourced to an agency or just the manpower required is outsourced and space and equipment is provided by the hospital.

Other support services include engineering (civil, electrical, plumbing, public health work), medical gases and bio-medical engineering. In a few hospitals there is one engineer to manage the services and the maintenance and repair is outsourced where as in others the department is fully managed by the regular hospital staff. Each hospital has one civil and one bio-medical engineer who are in turn supported by ITI/diploma holder technicians in each category.

The security services are outsourced by the hospitals but are managed by a regular Chief Security cum Fire Safety Officer. He or she ensures the proper performance of the security guards and undertake frequent rounds of the hospital.

Recommendations

As a result of the study following clinical and non-clinical services are suggested for district level hospitals of Haryana.

1. Clinical Functions

- ☆ OPD
- ☆ IPD
- ☆ Emergency
- ☆ Operation theater
- ☆ ICU/SNCU
- ☆ Labour room
- ☆ Blood bank
- ☆ High dependency units
- ☆ Dialysis
- ☆ Ambulance
- ☆ Physiotherapy
- ☆ Ayush

☆ ICTC

☆ De-adddiction centre

2. Diagnostic Services

☆ Laboratory

☆ Radiology

3. Non-Clinical Functions

☆ Medical records

☆ Central sterile supply department

☆ Mortuary

☆ Housekeeping and gda

☆ Laundary and linen store

☆ Horticulture

☆ Security and fire safety

☆ Kitchen

☆ Pest control

☆ Bio-medical waste

☆ Griveance readressal

4. Engineering Services

☆ Medical gases

☆ Plumbing

☆ Electric work

☆ Bio-medical equipment

☆ Non bio-medical equipment/machinary (lifts, generators etc)

☆ Sanitation

☆ Public health

☆ Maintenance and repairs of buildings

5. Purchase and Stores

☆ Medical stores/pharmacy

☆ General stores

6. Finance

7. HMIS/IT

8. Human Resource Management (HRM)

HRM basically involves following functions

☆ Hiring of contractual employees

☆ Handling duty rosters, vip duties, leave records, training records

☆ Maintaining personal files

☆ Capacity building of various levels of staff by regular training need assessment

☆ Making job descriptions and job specifications

☆ Performance appraisals/ACR

☆ Handling public relations like liasioning with press

☆ Grievance redressal

☆ Monitoring patient and employee satisfaction

☆ Handling legal matters

Proposed Administrative Strucutre for District Hospitals

Administrative Structure for Government Hospitals, Haryana:

1. There is need to strengthen the office of PMO in order to assist him/her in administrative work.

2. PMO should be assisted by two deputy Medical Superintendents (DMS-1 and DMS-2) and a Quality Manager (QM) in 200 beded hospitals, whereas in 100 beded hospitals two DMS can be posted and work of Quality Manager can also be looked after by them.

3. DMS-1 would look after clinical services and would be an MBBS with MHA, whereas DMS-2 would look after non-clinical services and would be MBBS/BDS/BAMS with MHA. Quality Manager would look after all the quality Improvement activities and would be responsible for co-ordination and implementation of various processes related to NABH accreditation.

4. Regular HCMS doctors with requisite qualifications would be posted as DMS.

5. Two supervisors each may be hired by Government on contract to assist DMS-1 and DMS-2. These supervisors in turn would be assisted by outsourced staff.

6. **DMS-1** would directly look after clinical services, diagnostics (Lab, Radiology), purchase (Drugs, Equipment) and Medical records. He would be assisted by computer assistant and Medical Record Technician. In addition he would also look after IT/HMIS/Registration, grievance redressal, pharmacy, ambulance services and CSSD with the help of two supervisors and outsourced staff.

7. **DMS-2** would look after housekeeping, laundry, bio-medical waste, infection control activities, sanitation, kitchen, horticulture, security and fire safety, bio-medical engineering, civil works and medical gases.

8. A retired **JE/SDO** would be hired to work under DMS-2 who would also be assisted by Bio-Medical Engineer, Chief Security Officer/Chief Fire Officer (CSO/CFO) and two supervisors.

Administrative Structure for District Hospitals

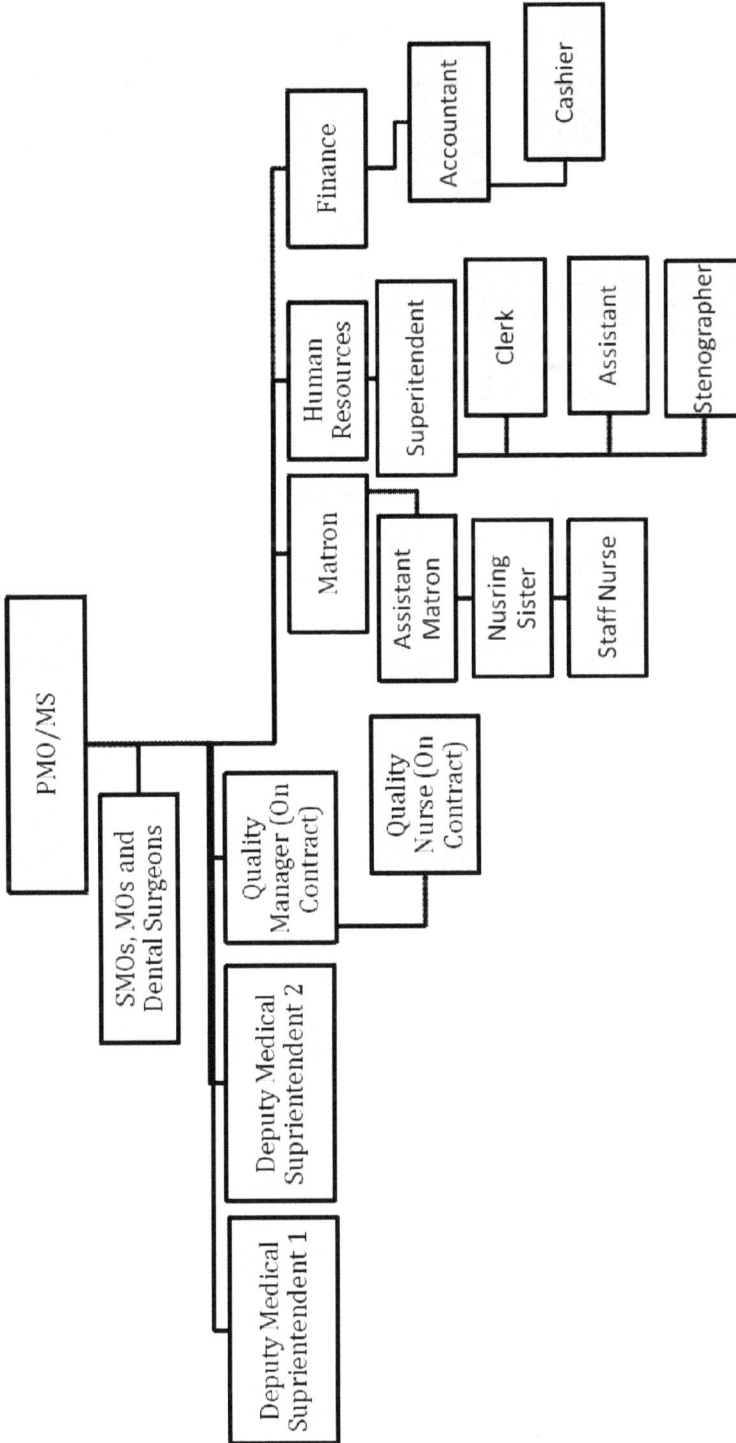

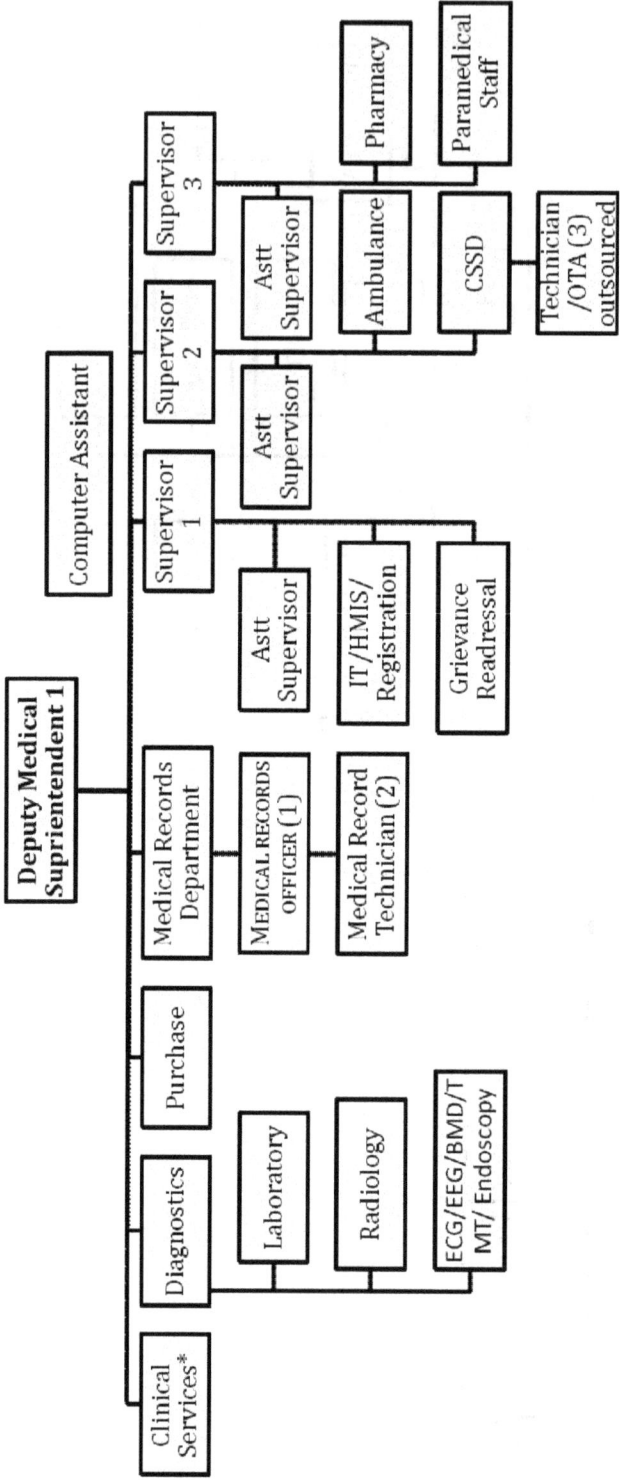

*Out patient department (OPD), Inpatient department, Operation Theater, Emergency, Labour Room, Intensive Care Units, Sick Neonate Care Unit, Dental Services, Blood Bank, High Dependency Unit, Dialysis, Ambulance, Physiotherapy, De-addiction Centre, ICTC, AYUSH

Deputy Medical Suprientendent 2

- Supervisor 1
  - Astt Supervisor
- Supervisor 2
  - Astt Supervisor
    - Laundry outsourced
    - Bio Medical Waste (Outsourced)
    - Infection Control Practices
  - Kitchen*
  - General Store
  - Horticulture
- Chief Security Officer + Fire Safety Officer
  - 02 Supervisors
  - Security Guards (Outsourced)
- Bio Medical Engineer
  - Jr. Engineer
  - Technician
  - Bio Medical Equipment
  - Medical Gases**
  - Non Bio Medical Equipment (Machinery)
- JE/SDO***
  - Electrical Work
  - Civil Works
  - Plumbing
  - Public Health
- Supervisor 3
  - Astt Supervisor
    - Housekeeping & GDA (Outsourced)
    - Sanitation (Outsourced)
    - Pest Control

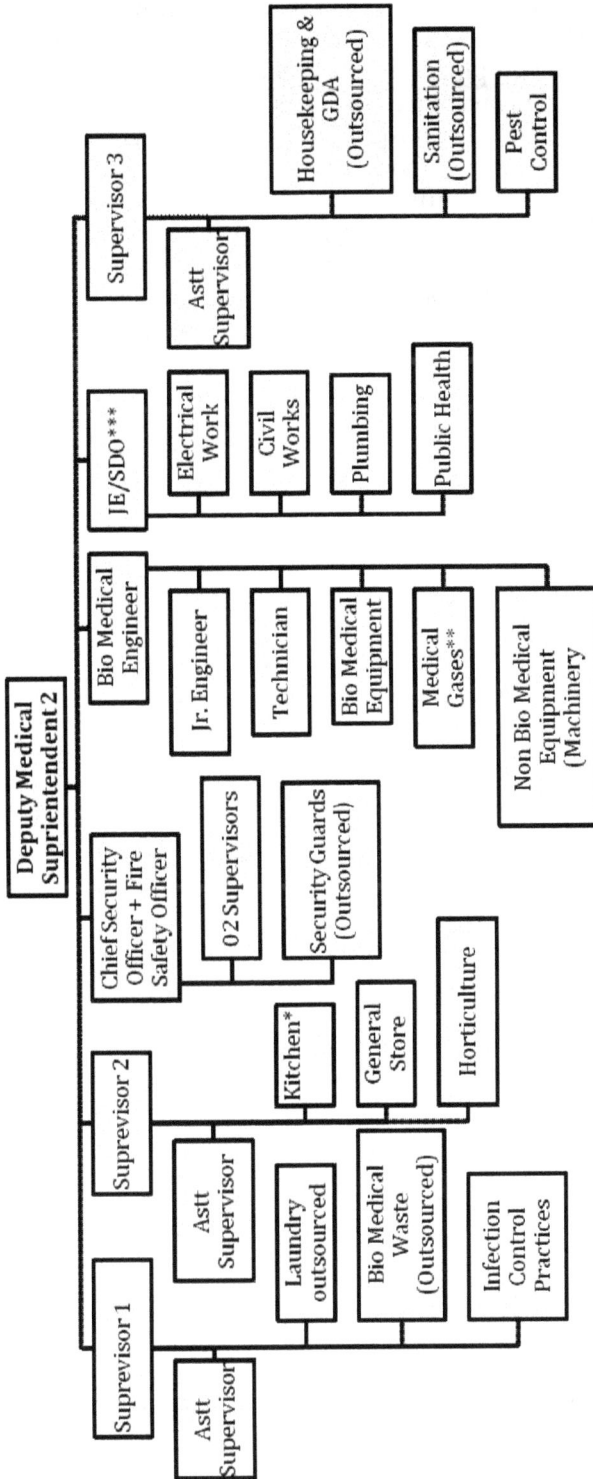

* Kitchen – 03 Cooks, 03 Helpers, 02 Sweepers (Regular), 8 Servers (2 per floor x 2 floors x 2 shifts)

** Medical Gases – 04 Technicians with ITI (contract/outsourced)

*** JE/SDO – Helped by 07 technicians electrical (03), civil work (01) and public health work (03) (contract/outsourced)

Class IV (Sweeper, security guard, chowkidar, ward servant, dressers, mali, dhobi etc.) – 81 through outsourcing.

9. JE/SDO would look after electrical work, civil work, plumbing and public health with the help of outsourced staff. Bio-Medical Engineer would look after Bio-Medical equipment, Non Bio-Medical equipment and Medical gases with the help of outsourced staff. CSO/CFO would be responsible for the security systems of the hospital and would also ensure fire safety of building and conduction of mock drills etc. and would supervise security guards hired through outsourcing Supervisor-1 would look after Housekeeping and GDA, Laundry, Bio-Medical waste and sanitation services with the help of outsourced staff. He would also look after infection control practices. Supervisor-2 would look after Kitchen and Horticulture with the help of outsourced staff. He would also look after general stores.

Chapter 10

Recognition of Women Farmers at State Level through Institutional Strengthening and Advocacy: Learning from Evaluation Study in Uttar Pradesh

*Nemthianngai Guite[1], Samik Ghosh[2]
and Aniruddha Brahmachari[3]*

*[1]Assistant Professor of Social Work and Public Health,
Department of Social Work, Delhi University, Delhi
[2]Programme Coordinator; [3]Manager,
Monitoring Evaluation Learning, Oxfam India, New Delhi
E-mail: [1]guitemahal@gmail.com;
[2]samik@oxfamindia.org and samikghosh_crj@hotmail.com;
[3]aniruddha@oxfamindia.org*

ABSTRACT

The women farmers in Uttar Pradesh, like in other parts of the country, were considered only as a 'helping hand' for men and were denied any role in household decision making. They also did not possess any control over agricultural production and marketing process. The project initiated by Oxfam India supported by Ford Foundation in India for its partner Non Governmental Organization (NGO), Vinoba Sewa Ashram in 2008-09 adopted an integrated strategy to establish the identity of women as a farmer and tried to develop their economic leadership. The project since its inception tried to provide due recognition for women farmers in their socially unacknowledged but significant contribution towards agricultural production in Shahjahanpur and Pilibhit districts of Uttar Pradesh. Oxfam India in association with its partner NGO took an initiative to organize women of 1928 households in 41 villages into Self Help Groups (SHGs), linked them with banks for credit, organized training and exposure visits, set up demonstration plots and promoted treadle pumps to instill confidence and paved the road for the economic independence of the women farmers in this area. In total 912 women farmers who are engaged in vegetable production in 130 and 177 hectare of lands in Shahjahanpur and Pilibhit districts respectively are involved with the project. The project has been successful with regard to enhancing food and nutrition security of women and children by capacitating women farmers in low cost vegetable production, recognition of women farmers at state level through institutional strengthening and advocacy, and developing and strengthening power in the market for women vegetable producers.

The paper made an attempt to share the specific impacts and learnings, best practices of the experiences of women farmers' participation at the project. The paper also presents record observations, findings in an objective manner and made an effort to discuss observations factually and distill those into workable recommendations for the future.

Keywords*: Women farmers, Vegetable production, Self help groups, Empowerment, Food security, Sustainable livelihoods.*

Background

In India, the agriculture sector employs 79 per cent of all economically active women; they make up 33 per cent of the agriculture labor force and 48 per cent of the self-employed farmers. In the state of Uttar Pradesh, the percentage of women farmers is higher at 38 per cent. However, despite the significant role played by women in agricultural production, the contribution of women farmers has never been recognized neither socially nor legally. Steady trends of migration, especially from the poorer states such as Uttar Pradesh, Bihar, and Uttarakhand; result in a further expansion of the role of women in the food system with limited control over land. Women's limited access to land not only prevents them from accessing government schemes and programmes for farmers; but also deprives them from accessing institutional credit; and limits their farm productivity. The districts of Shahjahanpur and Pilibhit traditionally fall within the vegetable production area in Uttar Pradesh. However, vegetable productivity in this region is lower than the national average because farmers are not aware about improved agricultural practices; and has limited access to credit, agricultural inputs, as well as the market. FAO has initiated number of programmes in partnership with NGOs to improve women's access to extension in agriculture through training, information and access to inputs and services. With these issues in mind, Oxfam India in partnership with local organization Vinoba Sewa Ashram (VSA) began working in both the districts in 2008, with a focus on building the economic leadership of women farmers by enhancing income from vegetable farming.

About Oxfam India

Oxfam India, a fully independent Indian organization (with Indian staff and an Indian Board) is a member of a global confederation of 17 Oxfams. The Oxfams are rights-based organizations that fight poverty and injustice by linking grassroots programming (through partner NGOs) to local, national and global advocacy and policy-making. All of Oxfam's work is framed by our commitment to five broad rights-based aims: the right to a sustainable livelihood, the right to basic social services, the right to life and security, the right to be heard and the right to equality: gender and diversity. (www.oxfamindia.org)

About VinobaSewa Ashram

Vinoba Sewa Ashram was founded in 1980 by Acharya Vinoba Bhave, at Brahma Vidya Mandir, Panwar, Wardha in Maharashtra. The guiding force behind VSA was noted Sarvodaya leader Nirmala Desphpande. The founders Vimla Behan and Ramesh Bhaiya chose an under-privileged area, 11 Km from Shahjahanpur towards Sitapur on Delhi-Lucknow highway for setting up the Ashram headquarters. Recognizing village as the base of development, Vinoba Sewa Ashram has made considerable progress during its 29 years of journey of service through a series of interventions in health facilities for rural women, education for their children and means of income generation for the family.

About the Project

The project in Shahjahanpur district and Pilibhit district was initiated in 2008-09 by Oxfam India in collaboration with partner NGO Vinoba Sewa Ashram. Despite the very significant role in agricultural production, women contribution has never been acknowledged, recognized and rewarded socially and legally due to prevalent negative mindset in the family and society. They are largely considered to be "a helping hand" of men and supposed to play supportive roles and hence deprived from decision-making and do not have control over agricultural production as well as marketing process. Women farmers are out of the mainstream of agriculture development, even agriculture extension program do not address the need of the women farmers instead they are discouraged due to typical nature of certain conditions imposed by the banks and other lending institutions. Women farmers are virtually "debarred" from availing credit facilities. In order to address these issues, Oxfam India launched its state level campaign to provide the due credit to women farmers, their identity as farmers.

The project covers a total of 1928 households in 41 villages of Shahjahanpur and Pilibhit (912 women farmers, covering 130 and 177 hectare of land for vegetable cultivation in Shahjahanpur and Pilibhit respectively). In Pilibhit district all the 16 villages comprise of Bengali farmers migrated from Bangladesh in 1971. They were dependent on the little land allocated to them by the government. In Shahjahanpur, the project is focused towards the betterment of women from Dalit, Muslim and Other Backward Caste from 25 villages.

Table 10.1: Project Coverage

District	Household	Village	Hectare of Land
Shahjahanpur	1016	25	130
Pilibhit	912	16	177
Total	1928	41	307

Key Issues Concerning the Project

Despite the fertile land, the status of farmers depending on it for livelihood is not so well. There has been several issues were identified that better be addressed to strengthen the livelihood of the farmers in the region:

☆ Land holding is in the name of men, which deprives women farmers from accessing credit and benefits of other government schemes.

☆ Large number of landless families. They cultivate cash crops or vegetables on land taken on lease at unjust conditions. They also do not have access to institutional credit and other government facilities.

☆ Farmers are not aware of improved agricultural practices resulting into inappropriate use of seed, manure, fertilizers and pesticides. Vegetable productivity of the area is lower than national average, which is mostly due to improper package of practices.

☆ Limited access to inputs quality like seed, fertilizers and pesticides.

☆ Inadequate access to credit through institutional channels.

☆ Farmers are unorganized and hence not having their voice in the market.

☆ Government extension services are not well accessible to farmers and untimely.

Locale of the Project Districts

The district of Pilibhit is located in the state of Uttar Pradesh. The district, in the central northwest of the state on the border of Nepal, extends over 3499 square kilometres between 28° 06' N and 28° 53' N, 79° 37' E and 80° 27' E.

On the Ganga Plain, it consists of a belt of marshy Tarai in the north sloping south at 189 metres into flat alluvial plain. Many parallel streams drain towards the River Ramganga. Rainfall averages 1276 millimeters supporting tropical moist deciduous forests along the rivers. 78 per cent of the area is cultivated, 55 per cent of this being irrigated mainly by 6 Sarda Canal in the east and bore wells. Agriculture is very less diversified in the district of Pilibhit. Mainly rice, wheat, pulses, sugar cane

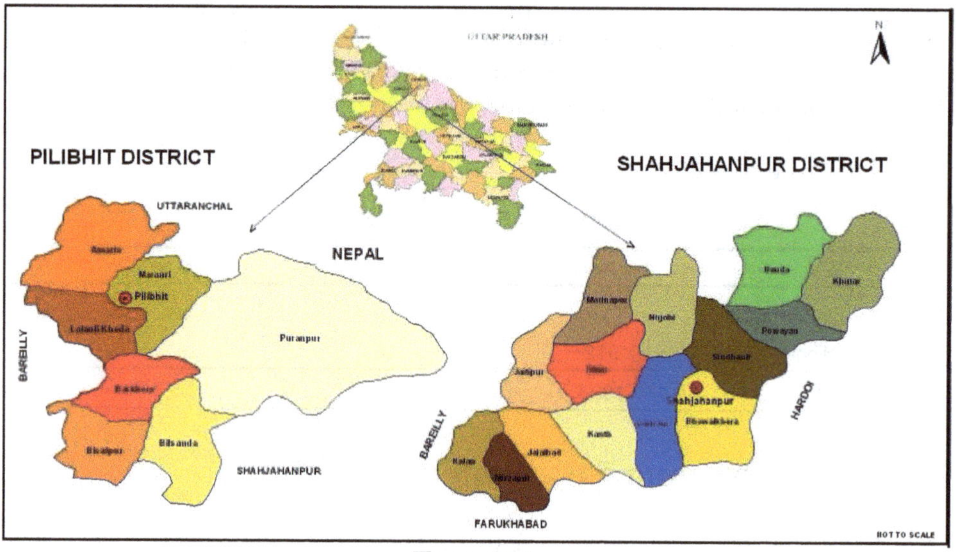

Figure 10.1

and oilseeds are grown. According to the 2011 Census, Pilibhit district has a total population of 2037225 and a sex-ratio of 889 males for every 1000 males, while the female literacy is 50 per cent.

Shahjahanpur is also a district in the state of Uttar Pradesh. The district, in the central northwest of the state, extends over 4574 square kilometers between 27° 28' N and 28° 28' N, 79° 17' E and 80° 23' E. On the Ganga Plain the district undulates at 150 metres, drained by parallel streams flowing south from Tarai in the north of the district to River Ganga. The Ramganga crosses the southwest corner to join Ganga, the south border, in the southeast, east of Ramganga are sandy ridges. It is a flood prone area Rainfall averages 1056 millimeters supporting patches of tropical moist deciduous forest in the northeast. Wheat, rice, maize, pulses, sugar cane, tobacco, potato, groundnuts and mango are grown in the district. According to the 2011 Census, Shahjahanpur district has a total population of 3006538 and a sex-ratio of 872 females for every 1000 males, while the female literacy is 49.57 per cent.

Baseline Learning: Strategy Formulated Based on Certain Parameters

A baseline study was conducted on the nutritional status of women vegetable producers in the project area. Some of the key findings that emerged were:

- ☆ Most of the women were homemakers, and worked on their own fields as unpaid labour.
- ☆ The average income per household per month was very low at Rs. 5120.
- ☆ 9 per cent of the total population was suffering from malnutrition.
- ☆ 73.2 per cent of the women reported that they cook either vegetable or pulses and not both; the intake of pulses was twice or thrice weekly.
- ☆ Cereals either in form of roti or rice constituted a major part of the daily diet.

End Term Evaluation

This evaluation was commissioned for learning purpose by Oxfam India with support from Ford Foundation in India for its partner organization VSA's vegetable farming project. It was undertaken to assess the success and impact of measures adopted under the project, to enhance the capacity and skills of women vegetable farmers in sustainable farming practices. At the same time it tried to document the problems and difficulties faced by women vegetable farmers in challenging the negativity prevalent in the family and society that deprived them the rightful recognition of their hard work. The study also intended to discover the problems and challenges faced by NGO and government officials in the implementation of the project.

The study entails a mixed methodology, as it aims to present the impact of program in terms of indicators with the help of interpretative and quantitative techniques.

Methodology

Purposive non-probability sampling was undertaken on a key set of stakeholders which included women vegetable farmers, NGO and government officials. Only those women farmers were selected who are currently a member of Vegetable Producer Groups (VPGs) formed under the SWFSLVP project. The total number of respondents selected for the present study as following:

☆ Women Vegetable Farmers: The size of sample taken for the women vegetable farmers were 100 (60 from Shahjahapur and 40 from Pilibhit) in all the clusters combined.

☆ NGO Officials: The total sample size of NGO officials was 8.

☆ Government Officials: The total sample size of government officials involved in the implementation of the project was 5.

Mostly primary data was used for the analysis and the tools used for data collection were interview schedules, FGD guides and case studies. Data was coded and organized into a codebook. Coded data was tabulated and transferred to a master table for analysis. Transcripts were prepared of the responses by women vegetable farmers, recorded for the purpose of case studies. These transcripts were translated from Hindi language to English and used for the study verbatim.

The main aim for carrying out the FGD was to understand the nature of participation of women vegetable farmers in strengthening women farmers for sustainable low-cost vegetable production (SWFSLVP) and the involvement of VSA in the program. It was conducted to explore and assess the diverse points ranging from land ownership pattern to income, SHGs, capacity building trainings, marketing business development issues and strategies.

Under the study **12** case studies have been recorded to explore and understand complex issues pertaining to transformations in the lives of women vegetable farmers after becoming part of the program. It also tried to represent on women economic leadership and changes occurring in their decision-making ability. It also tried to ascertain the challenges faced by them, their future plans and aspirations.

Integration of Women in the Development Projects: Evidence from the Field Based Evaluation Study

1. Capacitating Women Farmers in Low Cost Vegetable Farming

The study found that the project was successful in building capacity of women vegetable farmers for low-cost vegetable farming. Most of them were trained in low cost vegetable production. Some of them have undertaken exposure visits facilitated by VSA at several Agricultural research and training institutes in Pantnagar, Pusa Delhi, Varanasi, Lucknow, etc., thereby increasing the awareness levels of most of the farmers about organic manure, compost, insecticides and pesticides. The project has been successful in providing training on making Narayan Devrao Pandhari Pande (NADEP) compost, vermi compost pits, conducting exposure visits, and setting up demonstration plots and organic Package of practices (POP) manual provided to the beneficiaries. The strategy adopted for building capacity of the women farmers by giving them trainings and exposure to techniques of low cost vegetable production reflected the pathway of own production to consumption.

Table 10.2: Activities Conducted to Capacitate Women Farmers

Objectives 1	Activities
Enhance the food and nutrition security of women and children by capacitating women farmers in low cost vegetable production.	Capacity building on low cost production technology *i.e.*, green manuring, FYM based Fertilizer application, Bio pesticide application (IPM)
	Exposure visit to Vegetable Research Institute, Varanasi; Green Farm, Indore and Inter village visits
	Demo Plots for Experimental demonstration
	Meetings of District Advisory Committee (DAC)
	Conduct a study on the food and nutrition of the women in the project area and a baseline survey on the vegetable production
	Promoting Seed Bank
	Promoting the organic Manure preparation
	Green Manure

Practical and participatory on-farm demonstrations of sustainable cultivation techniques and practices have greater learning impact among farmers than conceptual classroom trainings by experts. The demonstration farms established under the project were optimally used as the learning sites for farmers facilitated by subject matter specialists from government agriculture extension department (Krishi Vigyan Kendras) and the project staff. The demonstrations pertaining to selection of appropriate plant varieties, soil nutrient management, sowing methods, integrated pest management were able to generate enthusiasm and learning and adoption interests amongst vegetable farmers even from non project villages.

2. Institutional Strengthening and Advocacy

Engaging with men and gender sensitization is critical for knowledge transfer and increased decision making role: The prevailing rural patriarchal system has been a major hindrance towards participation in training (especially residential) and accessing markets. It takes a lot of efforts by the field staff and women farmer leaders to convince male members in the families and even technical experts from government extension departments to ensure women vegetable grower participate in capacity strengthening processes such as training and demonstration sessions, exposure visits, interface with government officials, extension workers or traders. The partner organization also being the regional lead agency of statewide women farmer's campaign has made significant effort towards sensitizing farmers, government extension workers and officials towards women's role and contribution in farming that has paved the way for women vegetable growers to break the stereotype and establish themselves as farmers.

However, the project was also able to strengthen institutions to give due recognition to women vegetable farmers. It has successfully managed to form Vegetable Producer Groups (VPGs) in all the villages and encourage women to begin vegetable farming using low-cost sustainable technologies. These VPGs are running very effectively, meetings are held regularly, and members also attend these meetings regularly. In a step to further strengthen women vegetable farmers of Pilibhit and Shahjahanpur district, Oxfam India in collaboration with Margdarshak NGO has been able to successfully established federations of women vegetable farmers associated with the project. The main idea behind formation of such federations was to build strong linkages between producer and the market.

Table 10.3: Activities Undertaken to Strengthen Institution and Advocacy

Objectives 2	Activities
Recognition of women farmers at state level through institutional strengthening and advocacy	Inter linking the SHG's into Clusters and formation of the federation
	Development linkages with bank
	Development linkage with government scheme linkages
	Local cadre development through training @ Rs. 500/month as stipend for 24 months (*farmer field school concept*)
	State level public hearing on issues of women farmers
	Documentation of learning/success stories and dissemination

In 2013, first women vegetable farmer federation was established in the region. Around 50 per cent of the women vegetable farmers associated with VPGs under the project have become part of these federations. Currently, there are total 1850 members in the federations. This strategy of the project further reflected the pathway of mainstreaming women in the development process by enabling them to control food prices and purchase. Through the VPGs model approximately 2500 women farmers have been organized. They are encouraged for regular savings and credit activity along with their main role of sustainable farming. It is a step towards institution building for better bargaining power and reduced dependence on moneylenders. The number of VPGs formed under the project is 250, spanning across 21 villages.

Table 10.4: Block Wise Distribution of VPGs

Districts	Blocks	Village	No. of VPGs
Shahjahanpur	Bahawalkhera	Bartara	50
Shahjahanpur	Bahawalkhera	Mukrampur	50
Shahjahanpur	Jalalabad	Jalalabad	50
Pilibhit	Lalaurikhera	Lalaurikhera	50
Pilibhit	Puranpur	Ramnagara	50
Total			250

Facilitating convergence and leveraging resources from government schemes is a good motivation and road to sustainability. Creating awareness about relevant government schemes and programmes and helping women farmers to avail the entitlements and benefits has been helpful to build their confidence as well as establish their worth within the family and community. Linkages and leveraging resources with relevant government schemes such as Rashtriya Krishi Vikas Yojana (RKVY) and Green Revolution Scheme has enabled resource poor women farmers to access agriculture inputs such as seeds, fertilizers, tools and implements. These linkages are also critical to ensure sustainability of the project interventions and institutions created.

Recognition and appreciation helps assert the identity and respect locally and within family. Two women farmers have been nominated in the district committee of District Agriculture Technology Management Agency (ATMA), another women vegetable grower Leelavati was invited by Lucknow Centre of the National Television Channel (Doordarshan) to share her farming experience in popular agriculture programme *Krishi Darshan*. Several women vegetable growers from the project area were felicitated at the local, state and national level at the Women Farmers Felicitation events organized by Oxfam India. The recognition and appreciation has helped these women farmers establish their identity and leadership inspiring other women farmers from the area to reach out to them and seek their advice.

3. Economic Leadership and Strengthening Power

The project is encouraging and helping women vegetable farmers to realize their true potential. It is also facilitating in the transformation of their economic status, as well as that of their families. The activities that help in learning tricks of the trade are the key to create space in vegetable market. The traditional method of vegetable trading in conventional market is very secretive with sign language and has so far been a male bastion. Hence decoding and understanding the trade lingo is critical for women vegetable growers to enter and create space of their own in the market. The business development services interventions with women vegetable growers under the project strategically include sessions by traditional traders in the vegetable business besides the market experts and business planners to enable the women vegetable growers learn operational aspects of the vegetable markets.

The project intervention highlights that there is a significant increase in the land taken on lease by women vegetable farmers. While only 20 per cent of women farmers were doing farming on leased land in 2007-08, currently the percentage has gone up to 24. There is also a drastic change in the utilization of land for the purpose of vegetable farming in Shahjahanpur and Pilibhit districts. The farmers using 1-2 Bighas of land for vegetable cultivation has risen to 33 per cent from 17 per cent in 2007-08.

There are also significant changes taking place in the production of vegetables in the region. Currently more than 30 per cent of the women vegetable farmers are producing more than 50 quintals of vegetables, while the number stood at only 5 per cent in the year 2007-08. There is an enormous change in the pattern of vegetable production as well at end project evaluation, as 20 per cent of women vegetable farmers stated that vegetable farming is fetching them Rs 30,000 – 50,000 incomes annually while this figure stood at only 6 per cent in 2007-08.

In terms of family decision-making, 87 per cent of women vegetable farmers feel that after joining the project, they have more control over financial matters. As many as 85 per cent women vegetable farmers acknowledged that their ability to negotiate has immensely improved after becoming part of the project. Substantial number (71 per cent) of women vegetable farmers responded that credit has become more accessible to them due to the efforts made under the project. A significant number (71 per cent) of them have gained full market access, while 33 per cent of women vegetable farmers still lack easy and swift access to the market. Very high (95 per cent) proportion of women vegetable farmers reported that the level of their public participation has improved drastically. Similarly, majority (90 per cent) of them reported about an increase in the self-confidence.

Table 10.5: Activities Undertaken to Develop Economic Leadership

Objectives 3	Activities
Developing and strengthening power in the market for women vegetable producers	Training to leading women farmers on market dynamics
	Establishing 4 temporary storage cum information centres.
	Business Development Support on product diversification and market linkages
	Study of value chain on key products and product mix - vegetable diversity, seasonality and cash-flow and return

The strategy of the project adopted to build economic leadership reflected the pathway of women's control of income to resource allocation. Women's control of household income and their ability to influence household decision-making and household allocation of resources for food, health, and care comes only through achieving economic independence.

Conclusion

This project shows that women farmers from Shaharanpur and Pilibhit were effectively integrated into the development programmes and schemes of the government. Their active participation combined with the trainings and exposure provided under the activities of the project further increased their confidence level. This project is a powerful combination of pathways to achieve gender equality by recognizing the strength of women and providing them the opportunity and platform to prove their capability. The process of intervention and activities initiated are leading to economic leadership outcomes but is not unique in its application. What sets the project apart is its simultaneous focus on local market access through institutional strengthening to enhancing nutritional status of women and children by empowering women to have economic independence necessary to strengthen power in decision-making process. The activity oriented its intervention toward low cost vegetable production, promoting indigenous methods and techniques to increase vegetable production, encouraging dietary diversity, and generating revenue. At the same time, the project has worked to create a system—the VPGs network—that can provide improved seeds, extension services, agricultural services, and nutritional products. In addition, the importance of VPGs in empowering women in agriculture cannot be overstated. The project provides important lessons regarding the broader institutional linkages that ensure that gender-development integration can be effectively implemented. The project also consults with, and draws extensively on, government agricultural research facilities and both regional and local experts. The project has reached out on both sides of the agriculture – nutrition equation to ensure that its interventions are as coordinated as possible. The women farmer centric approach makes it possible to increase the impact of current initiatives, which aim to reverse the course of the major developmental issues.

Acknowledgement

The opinions expressed in this paper are those of the author(s) and do not necessarily reflect those of Oxfam. The paper is based on meta-reviews and endline evaluation study conducted by Oxfam India during the project phase out. A word of acknowledgement for Ford Foundation in India for supporting this project. The authors acknowledge the concerned regional team of Oxfam India based in Lucknow (UP) and Economic Justice team based in Delhi for extending valuable inputs and assistance. Sincere thanks and appreciation goes to the partner organization Vinobha Sewa Ashram in Uttar Pradesh for implementing this project. This evaluation would not have been possible without their support and reflections. A special thanks to the communities who shared information and participated whole heartedly in the evaluation process.

References

Apthorpe, Raymond (2000). 'Kosovo Humanitarian Programme Evaluations: towards synthesis meta-analysis and sixteen propositions for discussion' London: ALNAP www. alna~.orQ

Aubel, Judy (1999). Participatory Program Evaluation Manual: including program stakeholders in the evaluation process 2nd edition Baltimore: Catholic Relief Services

AusAid (1997). 'Monitoring and Evaluation Capacity Building Study'

Ayas, Karen and Zeniuk, Nick (2001). 'Project-based Learning: building communities of reflective practitioners' Management Learning 32:1: 61-76

Baker, Judy L. (2000). Evaluating the Impact of Development Projects on Poverty: a handbook for practitioners Washington DC: World Bank

Barnard, Geoff and Cameron, Catherine (2000). 'Evaluation Feedback for Effective Learning and Accountability: synthesis report' Brighton: Institute of Development Studies www.ids.ac.uklefeIaI

Aubel, Judi. Participatory Program Evaluation: A Manual for Involving Program Stakeholders in the Evaluation Process. Catholic Relief Services, Senegal, 1993.

Anayanwn, C. N. "The Technique of Participatory Research in Community Development", *Community Development Journal*, January 1988, vol. 23:11-15.

Armstrong, John and Michael Key. "Evaluation, Change and Community Work", *Community Development Journal*, October 1979, Vol. 14:210-223.

Banlina, F. T. and Ly Tung. "Farm Experiences of Wealth Ranking in the Philippines: Different Farmers Have Different Needs". International Institute for Environment and Development. "Special Issues on Application of WealthRanking". RRA Notes, no. 15, May 1992, pp. 48-51.

Campos, Juanita Diane. Towards Participatory Evaluation: An Inquiry into Post Training Experiences of Guatemalan Community Development Workers. Dissertation submitted to the University of Massachusetts, School of Education, May 1990.

Chambers, Robert. "Participatory Rural Appraisal (PRA): Analysis of Experience", in World Development, Vol. 22, No. 9, 1994:1253-1268.

Coupal, Françoise. "Participatory Project Design: Its Implications for Evaluation. A Case Study from El Salvador". Paper presented at the 1995 Evaluation Conference, 1-5 November, Vancouver, British Columbia.

Farrington, John and Adrienne Martin. "Farmer Participation in Agricultural Research: A Review of Concepts and Practices", Agricultural Administration Occasional Paper no. 9, London, 1988.

Fernandez, Maria. *Participatory-Action Research and the Farming Systems Approach with Highland Peasants.* Colombia, Department of Sociology, University of Missouri.

International Institute for Environment and Development. "Special Issues on Application of Wealth Ranking". RRA Notes, no. 15, May 1992.

Kassam, Yusuf. "The Combined Use of Participatory and Survey Methodology in Evaluating Development Projects. A Case Study of a Rural Development Project in Bangladesh". September 1995. Paper presented at the 1995 Evaluation Conference, 1-5 November, Vancouver, British Columbia.

Marsden, David, Peter Oakley and Brian Pratt. *Measuring the Process: Guidelines for Evaluating Social Development.* Oxford, 1994.

Pretty, Jules, Irene Gujit, Ian Scoones and John Thompson. *Participatory Learning and Action: A Trainer's Guide.* London, International Institute for Environment and Development, 1995.

Section III

GENDER AND EVALUATION

Chapter 11

An Evaluative Analysis of Unemployment in India

Farha Anis

Niti Aayog
New Delhi
E-mail: farha.anis13@gmail.com

ABSTRACT

The last decade witnessed a diverging trend between the structure of output and the structure of employment in India, particularly when GDP growth has increased significantly. This article tries to capture the diverging trends of employment vis-a-vis GDP growth over the two periods (*i.e.* 2004-05 to 2009-10 and post 2009-10 period). It also tries to analyze the unemployment trend (both gender-wise and sector –wise i.e rural-urban) since 2004-05 and also explores the regional variations in unemployment level in India over the period 2009-10 and 2011-12, based on study of various rounds of National Sample Survey unit level data. The major findings emerged from article are: the period from 2004-05 to 2009-10 witnessed a period of jobless growth. During this period, growth in GDP were not in proportion to growth in total labour force. Notwithstanding the decline in total LFPR, rural male LFPR witnessed some improvement during this period. After the disappointing performance between 2004-05 to 2009-10, the Indian labour market showed some improvement in post 2009-10 period. The urban LFPR increased on account of increase in urban male and female LFPR while rural LFPR declined, resulting in declining the overall LFPR during 2009-10 to 2011-12. The article also figure out the difference in the unemployment rate, based on usual status adjusted and current daily status. Moreover, the state-wise analysis of unemployment shows that the unemployment rate (UR) has increased for more than 50 per cent of the states from 2009-10 to 2011-12. On an average, Southern and western region had lower level of unemployment rate in 2011-12 than rest of the country. Some of the north-eastern states and islands/UTs had significantly high URs. Keeping this in mind, the govt. needs to create an enabling environment for employment-led growth particularly in northern region. GDP growth should take into account the sufficient employment level. In addition, govt. needs to enhance the scope of unemployment statistics for better research in this area.

Keywords: *Unemployment, Economic growth, LFPR, WFPR.*

An inclusive growth cannot be achieved without generating sufficient employment level.India witnessed a strong economic growth over the last decade, which has not translated into higher employment level. This article tries to explore the unemployment trends with GDP growth. It also tries to capture the gender-wise unemployment trends in rural and urban areas. Unfortunately, unlike advanced economies such as the United States, unemployment statistics for India are not available on a monthly basis. The quinquennial employment and unemployment survey of the National Sample Survey (NSS) is the primary source of data on various indicators of the labour force at the national and State levels. Researchers label this as the thick round (number of households surveyed in the first stage is more in the quinquennial rounds than the "thin" annual rounds). For the first time in 2011–12, the "thick" survey was carried out within two years of the previous one. The focus of this article is to analyse the unemployment rates for the three available periods: 2004–05, 2009–10 and 2011–12, for any discernible impact of the economic trends on unemployment. The Gross Domestic Product (GDP) growth rate for this period is shown in Table 11.1, with India having seen both the high and the low in a nine-year period.

Table 11.1: Growth Rate of GDP Factor Cost and Market Price (2004–05 price), 2004–05 to 2013–14

Years	GDP at Factor Cost	GDP at Market Price
2004–05	7.1	7.9
2005–06	9.5	9.3
2006–07	9.6	9.3
2007–08	9.3	9.8
2008–09	6.7	3.9
2009–10	8.6	8.5
2010–11	8.9	10.3
2011–12 (2nd RE)	6.7	6.6
2012–13 (1st RE)	4.5	4.7
2013-14 (PE)	4.7	5.0

Note: 2nd RE: Second Revised Estimate; 1st RE: First Revised Estimate; PE: Provisional Estimate.

Source: Central Statistical Office (CSO).

2004–05 TO 2009–10

During the period 2004-05 to 2009-10, India has had four years of consecutive growth rates above 8 per cent since 2005–06, except the year 2004-05 and 2008–09. Within a year of 2008–09 crisis, India showed strong recovery for two consecutive years. During the year 2004-05 to 2009-10, total labourforce and workforce increased from 468.7 million to 472.3 (0.8 per cent) and 462.5 million to 457.6 million (1.1 per cent) respectively. Theunemployment rate fell from 11.2 million to 9.8 million(12.5 per cent) during the same period. This period witnessed a jobless growth. Using NSS data, Chowdhury (2011) also shows that jobless growth took place during the year

2004–05 to 2009–10 *i.e.,* size of the workforce remained almost the same over the five years.

During the period, labour force participation rate (LFPR)[1] fell for all groups except rural men (which witnessed a minor increase). The same trend is witnessed in workforce participation rate (WFPR)[2] (Table 11.2). The decline in LFPR could be attributed to number of factors such as, changes in the demographic profile of the young population, rising enrolments in elementary and secondary schooling due to efforts of SarvaShikshaAbiyan (SSA) and Right to Education (RTE), declining child labour, fall in absolute number of rural female particularly from agriculture sector and absolute decline in manufacturing employment.Consequently, as people left the job market, unemployment rate (UR) fell from 23 per cent to 20 per cent over the period. Both rural and urban URs fell between 2004–05 and 2009–10. It fell for all the groups except rural male UR which remained the same in the two periods.

Table 11.2: Unemployment Rate (per 1,000 persons)
Usual Status (ps+ss), 2004–05, 2009–10 and 2011–12

	Rural			*Urban*			*Total (Rural Urban)*		
	Male	*Female*	*Persons*	*Male*	*Female*	*Persons*	*Male*	*Female*	*Persons*
2004-05	16	18	17	38	69	45	22	26	23
2009-10	16	16	16	28	57	34	20	23	20
2011-12	17	17	17	30	52	34	21	24	22

Note. 1. Unemployment Rate (UR) is defined as the number of persons/person-days unemployed per 1,000 persons/person-days in the labour force (which includes both employed and unemployed). Usual status adjusted UR: Unemployment rates for the reference period of 365 days, *i.e.,* in usual principal status (ps) approximates an indicator of chronically unemployed. Some of the persons categorized as unemployed according to the usual principal activity status might be working in a subsidiary capacity. Therefore, another estimate of the unemployed excluding those employed in a subsidiary capacity during the reference period can be derived. The former is called the usually unemployed according to the principal status (ps) and the latter is the usually unemployed excluding those employed in subsidiary status or usual status (ss) adjusted, *i.e.,* us (adjusted), which would conceptually be lower than the former.The current daily status (CDS) is based on the number of person-days unemployed on an average, on a day during the reference period of seven days preceding the date of survey. It gives average level of unemployment on a day during the survey year.

Source. NSSO 61st (July, 2004 to June 2005), 66th (July 2009 to June 2010) and 68th rounds (July 2011 to June 2012).

The unemployment rate based on usual status (adjusted), is based on the number ofpersons unemployed for a relatively long period during a reference period of 365 days excluding those employed in a subsidiary capacity during the reference period. While, theCDS approach is based on the daily activity pursued during each day of the reference week, therefore, it measures the extent of unemployment in terms of

---

1 LFPR: Number of persons employed + number of persons unemployed /total population*1000.

2 WFPR: number of persons employed/total population*1000.

person-days rather than unemployed persons. To an extent the unemployed days of persons who are usually employed are also taken into account in this case. In India, where large-scale seasonal unemployment exists, the CDS measure gives a better picture of the unemployment situation than the one relating to unemployed persons. Moreover, looking at such a high level of poverty, the poor people cannot afford to be unemployedin India over a long period of time. Therefore, CDS gives the better picture of unemployment, thus, this article analysis the unemployment rate based on CDS approach.

Table 11.3: Unemployment Rate (per 1,000 persons) Current Daily Status (CDS), 2004–05, 2009–10 and 2011–12

	Rural			Urban			Total (Rural Urban)		
	Male	Female	Persons	Male	Female	Persons	Male	Female	Persons
2004-05	80	87	82	75	116	83	78	92	82
2009-10	64	80	68	51	91	58	61	82	66
2011-12	55	62	57	49	80	55	53	66	56

Source: NSSO 61st (July, 2004 to June 2005), 66th (July 2009 to June 2010) and 68th rounds (July 2011 to June 2012).

Overall, the unemployment rate, according to CDS fell significant from 82 per cent in 2004-05 to 66 per cent in 2009-10 (19.5 per cent) as compared to unemployment rate based on usual status adjusted (13.0 per cent). It even decline for all the groups of households.

POST 2009–10

India's GDPgrowth (at factor cost) slowed to 6.7 per cent in 2011–12 from 8.9 per cent in 2010–11 (Table 11.1). It reached to 4.7 per cent in 2013-14. After a disappointing performance between 2004-05 and 2009-10, Indian labour market showed some improvement between 2009-10 and 2011-12. The total labour force increased from 472.3 million in 2009–10 to 483.8 million in 2011–12 (2.4 per cent). However, LFPR declined from 40 per cent in 2009–10 to 39.5 per cent in 2011–12. Rural employment declined on account ofdecline in both rural male and rural female employment. However, total urban LFPR has increased by 1.4 per cent on account of increase in urban female and urban male LFPR from 14.6 per cent 2009–10 to 15.5 per cent in 2011–12 and 55.9 per cent in 2009–10 to 56.3 per cent in 2011-12 respectively. The number of persons in the workforce increased from 462.5 million in 2009–10 to 472.9million in 2011–12, a growth of more than two per cent. Workforce participation rate (WFPR) declined from 39.2 per cent in 2009-10 to 38.6 per cent in 2011-12. Except rural females and males, urban males and urban females witnessed an increase in WFPR (from2009–10 to 2011-12).However, the number of unemployed measure increased from 9.8 million in 2009–10 to 10.8 million in 2011–12, an increase of 10.2 per cent.During the period, UR fell marginally from 20 per cent to 22 per cent.Urban male UR rose in 2011–12 over 2009–10 (Table 11.2). In contrast, the urban female UR fell a little in this period. In the rural areas, across genders, one sees a rise in the UR, albeit marginal. Overall, there has been a rise in the UR in 2011–12 over 2009–10.

**Table 11.3: State-wise Unemployment Rate (per 1,000 Persons)
Current Daily Status (CDS), 2009–10 and 2011–12**

Survey Year	July 2009–June 2010	July 2011–June 2012
Year	2009–10	2011–12
Andhra Pradesh	70	56
Arunachal Pradesh	19	24
Assam	65	55
Bihar	57	50
Chhattisgarh	34	56
Delhi	34	46
Goa	52	52
Gujarat	50	24
Haryana	55	45
Himachal Pradesh	44	23
Jammu and Kashmir	54	67
Jharkhand	75	39
Karnataka	42	36
Kerala	167	156
Madhya Pradesh	65	36
Maharashtra	63	40
Manipur	43	44
Meghalaya	15	12
Mizoram	24	33
Nagaland	150	241
Odisha	79	83
Punjab	65	49
Rajasthan	33	39
Sikkim	42	21
Tamil Nadu	117	93
Tripura	149	188
Uttarakhand	49	56
Uttar Pradesh	53	54
West Bengal	70	79
Andaman and Nicobar Islands	111	92
Chandigarh	89	65
Dadra and Nagar Haveli	50	4
Daman and Diu	34	2
Lakshadweep	167	206
Puducherry	143	95
All-India	66	56

Note. PS is principal status and SS is subsidiary status.

Source. NSSO 61st, 66th and 68th round.

The totalUR,according to CDS fell significant from 66 per cent in 2009-10 to 56 per cent in 2011-12. During the same period, it increasedaccording to usual status adjusted approach. It even decline for all the groups.

Overall, the unemployment rates, according to the current daily status (cds) approach, are higher than the rates obtained according to 'usual status' approach and 'weekly status' approach, thereby indicating a high degree of intermittent unemployment. This is mainly due to the absence of regular employment for many workers

State-wise Unemployment Trends

The analysis of state-wise unemployment rate reveals that there has been significant increase in unemployment level over the period (2009-10 to 2011-12). Spatially, the fall in the UR between 2009–10 and 2011–12 was uneven (Table 11.3). Many northeastern states namely, Arunachal Pradesh, Manipur, Nagaland, Mizoram, Tripurahad a significant rise in the UR. Assam and Meghalaya were the significant exception amongst the northeastern states.

URs have increased for more than fifty per cent of the states between 2009–10 and 2011–12. Dadra and Nagar Haveli, Daman and Diu, Meghalaya, Sikkim, Himachal Pradesh and Gujarat had the lowest unemployment rates in 2011–12 and Nagaland the highest. However, LFPR declined in Gujarat between 2009–10 and 2011–12 while it increased in Nagaland. On average, the southern and western regions had lower unemployment rates in 2011–12 than the rest of the country with significant exceptions like Lakshadweep, Goa and Kerala.

To sum up, urban females' LFPR works in a puzzling manner – falling in good times and rising in bad times. One may infer that urban females' earnings are viewed as supplementary income and therefore their LFPR rise during periods of slower economic growth. The unexpected silver lining of these uncertain times may be just that social conservatism may give way to hard economic realities.

Indian economy is beset with imbalances. LFPR has fallen while unemployment has risen between 2009–10 and 2011–12. There had been further fall in economic growth and rise in inflation in 2012 and 2013 and therefore one can intuitively forecast that it will have a further worsening impact on the job market.Employment is crucial in achieving growth with equity and pro-poor growth. The link between economic growth and employment is thus a process in which output growth induces an increase in productive and remunerative employment, which, in turn, leads to an increase in the incomes of the poor and a reduction in poverty. Fall in unemployment rates have been uneven spatially and worsen the signal jobless economic growth. Therefore, increasing the growth and emphasizing on patterns of economic growth which encourages creation of jobs shall be the focus of the new government. In addition, government needs to enhance the scope and quality of statistics on unemployment which shall be made available for better policy action in this regard.

References

1.   Central Statistical Office, Provisional esimates of Annual Income, 2013-14 and Quarterly estimates of Gross Domestic Product, 2013-14 http://mospi.nic.in/Mospi_New/upload/nad_pr_30may14.pdf

2.   Central Statistical Office, First Revised Estimates of National Income, Consumption Expenditure and Capital Formation, 2012-13 http://mospi.nic.in/Mospi_New/upload/nad_press_release_31jan14.pdf

3.   Chowdury, S. 2011. Employment in India: What does the latest data show? Economic and Political Weekly.46 (32). August 6. 23–26.

4.   Mehrotra S, et al. 2014. Explaining Employment Trends in the Indian Economy: 1993-94 to 2011-12. Economic and Political Weekly. August 9, VolxlIX no 32.

5.   National Sample Survey Organisation (NSSO), Key Indicators of Employment and Unemployment in India, 2009-10, 66[th]Round (July 2009-June 2010).

6.   National Sample Survey Organisation (NSSO), Key Indicators of Employment and Unemployment in India, 2011-12, 68th Round (July 2011-June 2012).

7.   National Sample Survey Organisation (NSSO), Employment and Unemployment Situation in India, 2004-05 (Part-1), 61st Round (July 2004 -June 2005).

8.   Rangarajan C, Seema, Vibeesh E M, 2014. Development in the Worforce between 2009-10 and 2011-12. Economic and Political Weekly. June 7, VolxlIX no 23.

Chapter 12

Review and Learning from an Endline Evaluation Commissioned by OXFAM India for the Programme "Promoting Violence-Free Lives for Women from Poor and Marginalized Communities in India"

Aniruddha Brahmachari[1], Samik Ghosh[2] and
Julie Thekkudan[3]

[1]*Manager – Monitoring Evaluation Learning*
[2]*Programme Coordinator – Monitoring Evaluation Learning*
[3]*Lead Specialist - Gender Justice*
Oxfam India, New Delhi
E-mail: [1]*aniruddha@oxfamindia.org;* [2]*samik@oxfamindia.org and*
samikghosh_crj@hotmail.com; [3]*julie@oxfamindia.org*

ABSTRACT

Violence Against Women (VAW) continues to be one of the most prevalent and least recognized human rights violations in the world. In India, this violence occurs in many forms - domestic violence, sexual assault, public humiliation or abuse, trafficking and honor killing. One in every two women in India experience violence in their daily life. Oxfam India, has major focus of working on gender issues that seeks to mainstream gender justice across all thematic portfolios. In addition, the Gender Justice theme of Oxfam India covers two programs: (a) Reduction of Violence Against Women (VAW), and (b) Political Empowerment of Women. The goals of two programs identified for Oxfam India are - reducing social acceptance of violence against women and; increased and effective representation of women in decision making forums in governance institutions. This paper has reviewed in first part about how the programme evaluation addressed various layers of violence exist within gamut of human perceptual beliefs, attitudes and practices and in the second part it is attempted to reflect on effective use of mixed-methods with vignettes within sphere of quasi-experimental evaluation research design was contextualized. The paper has documented evidence of impact and learning by documenting execution of development programming and about how mixed method approach of program evaluation can elaborate knowledge exchange.

Introduction

An analysis of development indices shows that nearly 70 per cent of the poorest and most marginalised in India are women. The Human Development Report (UNDP 2004) expressed concern for the status of women in India[1]; there is ample evidence that women from marginalised communities[2] experience further exclusions across different aspects of social development.[3]

More directly, gender-based violence at different levels has contributed to women's social exclusion and poverty. For instance, it has resulted in excessive mortality among women in India, whether as a foetus due to sex selective abortion or as a mother due to neglect and discrimination in access to health services and nutrition. According to Nobel laureate Dr. Amartya Sen, currently, 39.7 million women are 'missing' in India.[4] The situation will worsen with the drop in sex ratios, say experts. They fear greater violence against women and a further reduction in development indicators for them.

Women experiencing violence often are forced to remain silent and accept the violence as the formal justice institutions (police, courts) have failed to respond to women survivors and more towards vulnerable communities. Women from marginalised communities are even less likely to report violence or seek justice as they do not believe that they will get justice. This perception is borne out by reality as a National Campaign on Dalit Human Rights study (2006) showed that perpetrators were punished in only 1 per cent of cases.

Many a times, other institutions like counselling centres, short stay homes for women survivors are poorly resourced, inadequate for the population and often inaccessible. Support services offered by NGOs and other organisations are heavily burdened and few in number. Thus, there is a clear need not only to provide efficient and effective support mechanisms for women facing violence from all socio-economic

---

1   According to the UNDP report, women's life expectancy at birth is 64.4 years; adult female literacy rate is 46.4%; estimated earned income (PPP US $) is 1,442; maternal mortality ratio is 540 per 100,000 live births; infant mortality rate is 67 per 1,000 live births and only 9.3% seats in parliament held by women.

2   Marginalised groups are those who face discrimination and/or exclusion at a community, state or institutional level on the basis of their social identity. This marginalisation may be on the basis of their ethnicity, religious, caste, socio-economic status, disability, gender or sexual orientation, or other, contextualised, basis.

3   The literacy rate of is lower at just 23.8% with higher drop out rate of 53.96% among Dalit girls at the primary school level. Poverty and unemployment rates are higher for 94% of engaged the unorganized, self-employed sector (farm/wage workers, domestic helpers, etc.),

4   Klasen and Wink, 2004. 2011 (940 girls per 1000 boys)

backgrounds but also to create mechanisms at different levels for institutional accountability. Though there are some positive and progressive laws providing for time-bound justice for women, much needs to be done for effective implementation. There is a need to further engage government departments and judiciary to ensure effective entitlements by poor and marginalised women.

Gender-based violence at different levels in our society has contributed to women's social exclusion and poverty. For instance, it has resulted in excessive mortality among women in India, whether as a foetal loss due to sex selective abortion or as being mother due to neglect and face discrimination in access to health services and nutrition. Apart from this women in our society are excluded from many other amenities in their life. Under this backdrop OXFAM India has decided to focus on addressing violence against women (VAW) focusing on the 'domestic violence' as the key aspect of social exclusion and better implementation of the Protection of Women from Domestic Violence Act 2005. OXFAM India has further taken up the IPAP project with the objective of improving the status of the poorest and most marginalized women in India at identified states. The initiative aims to build upon the experience of work on ending violence against women undertaken by Oxfam India and its partner NGOs in different parts of the country. The project has been implemented since 2009 to 2014 in the states of Odisha, Andhra Pradesh, Gujarat, Uttar Pradesh and Uttarakhand through setting up and running of support centers and advocacy with government on implementation of the PWDVA act.

The initiative aims to build upon the experience of work on ending violence against women undertaken by Oxfam and its partners in different parts of the country. As envisaged within IPAP, the project is envisaged to impact at three levels – 1) focused outcomes at national and state level in policy implementation; 2) formal and non-formal institutional support mechanisms to survivors 3) broad-based community mobilization intervention for a fundamental shift in ideas, beliefs and practices of individuals and institutions that support and perpetuate violence against women.

The programme targeted all the actors that are involved in either perpetuating VAW or not fulfilling their mandate in prevention and relief, these being the individuals (perpetrators and survivors), families, communities, society and the state. Appropriate capacity building was done at all levels to work towards the larger goal of reducing social acceptance of VAW through addressing patriarchal and other interlinking discriminatory attitudes and practices based on caste, race, religion and bringing a positive change in the policy and programme environment that perpetuates its acceptance at an individual, community and institutional levels. All this would ensure that women from marginalised communities are able to overcome the multiple exclusions they face and are able to build an environment and support system that enables them to lead lives to the fullest potential.

Methodology

In 2010, a detailed baseline survey was conducted across the program area which reached out to 6275 respondents (3,055 men and 3,220 women aged 15-50 years) and 432 victims of women survivors faced gender based violence who had registered cases (with PO or police stations in the program area). The project has enabled systematic use of Monitoring Evaluation Learning (MEL) resources to document programme output and outcome. Over four years of implementation of the project life consensus made to capture evidence based learning for OXFAM India to measure the changes happened in patriarchal and other discriminatory social practices and belief system that perpetuate all violence against women. In addition attempts made how far the project have had captured policy makers and civil society response to address existing scenario on violence against women (VAW) focusing on the 'domestic violence' as the key aspect of social exclusion and level of institutional system sensitized and for implementation of the Protection of Women from Domestic Violence Act 2005. To measure IPAP programme outcome, endline evaluation study was commissioned referring to OXFAM Evaluation Policy. The study got ethical clearance from Institutional Ethics Review Board at the inception of the study. The Endline Evaluation criteria were based on the OECD – DAC criteria which are standard for international development evaluation since 1991. The questions that the evaluation intend to answer have been organized against the criteria of relevance, efficiency, effectiveness, sustainability and impact.

Evaluation Design

A **Quasi-Experimental (QE) Mixed-method (MM) design approach** was adopted for the study. Judicious mixes of quantitative and qualitative tools were used to carry out the study. Since a comparison group was not included in the baseline, **Difference-in-Difference** method could not be used. At best, **Propensity Score Matching was** used to create analysis across treatment and comparison arms on scientifically matched data sets.

The evaluation design cuts across the following stakeholder categories:

- ☆ Women who have registered VAW cases in the sample area
- ☆ Police personnel in police stations located in the sample area
- ☆ District level official in WCD department in the sample area
- ☆ POs (district level) in the sample area
- ☆ Civil society network members
- ☆ Youth activist/pressure group/volunteer in the sample area
- ☆ Counselors of counseling centers

The Table 12.1 describes the treatment of different elements for the evaluation.

Sampling Plan

The key objective of the evaluation was to measure and compare changes on increased knowledge on laws related to violence against women and legal and other support services available for those experiencing violence in men and women aged 15-50 years. To measure changes, which could have occurred due to intervention, sample size should be statistically adequate to identify and measure those changes. The sample size decision for detecting changes from baseline depends on the power

Table 12.1: Methodology Description

Objectives	Measurement Approach	Tools and Respondent Categories
Understand if men and women aged 15-50 years in the program districts are sensitized on VAW and have demonstrably increased their knowledge on laws related to violence against women and legal and other support services available for those experiencing violence	This objective is measured using a mix of Quantitative and Qualitative methodologies. While household interviews with men and women aged 15-50 years provides quantitative estimates to key programme indicators corresponding to the objective, community level focus group discussions provides qualitative insights and qualitative read out to the objective's metrics.	**a. Men and women aged 15-50 years** ☆ Household level respondent questionnaire ☆ FGD guidelines
Understand if communities in villages and towns in the program districts have pressure groups and/or youth activists and/or committees on VAW that are equipped with the knowledge of laws related VAW and support services to support women experiencing violence in seeking appropriate services	This objective primarily gets covered through qualitative means while there would be few corresponding quantitative variables included in the household level interviews with the men and women aged 15-50 years.	**a. Men and women aged 15-50 years** ☆ Household level respondent questionnaire ☆ FGD guidelines **b. Pressure group/activists/committee** ☆ Structured questionnaire
Understand whether Government officials in the relevant departments (Police, PO under DV act, Women and Child Development officer) are sensitised about the issue of VAW and have increased knowledge of the laws, rules and regulations related to VAW and support services and related budgetary allocation procedures to be followed for expenditure	Semi-structured tools would be used to conduct in-depth interviews with Government Officials in relevant departments. Semi-structured tool includes descriptive as well as quantitative questions.	**a. Government Officials in relevant departments** ☆ Semi-structured in depth discussion guideline
Understand the functioning of district level support institutions established in collaboration with police department in four states (Andhra Pradesh, Uttar Pradesh, Orissa and Gujarat).	Along with conducting in depth discussions with district officials the methodology also subsumes collection of secondary information/data on the district level support institutions. An in-depth discussion guideline is developed for conducting in depth discussions while a secondary data checklist would be used for the collecting secondary data.	**a. Government Officials in relevant departments** ☆ Semi-structured in depth discussion guideline ☆ Secondary data checklist

Contd...

Table 12.1–Contd...

Objectives	Measurement Approach	Tools and Respondent Categories
Understand if Targeted Education Institutions covered by We Can in 13 states adopt Violence against women prevention measures.	This objective is described using In-depth discussions with representatives of targeted education institutions. Secondary data from the targeted education institutions is collect using a secondary data checklist. Secondary data helps us create data points around what the institutions report with respect to adoption of violence against women prevention measures.	**a. Targeted education institutions** ☆ Semi-structured in depth discussion guideline ☆ Secondary data checklist
Understand whether We Can campaign is recognised by state and civil society organizations in 14 States as one of the key platforms addressing VAW issues	In depth discussions is conducted with representative civil society organizations in 14 states. Along with this evidences on measures taken by the states on recognizing "We Can Campaign" are collect. This is done through review of existing literature available with the Government and conducting in-depth discussions with key Government functionaries.	**a. Civil society organizations** ☆ Semi-structured in depth discussion guideline ☆ Literature review
Apart from the program planned strategies, a number of innovative initiatives were funded. The end line would try to make an inquiry into these initiatives also. These are mostly in terms of Joint Programming and Studies.	Develop case studies on innovative initiatives identified by the client.	Case studies

i.e., efficiency to detect and measure change, besides depending on level of statistical significance. The sample size required to assess change was computed based on following key parameters:

- ☆ Initial value of variable of interest (taking initial value from NFHS-III: ever-married women who have ever experienced spousal violence)

- ☆ Expected change programme was designed to make, which needs to be detected – robust enough to even detect a change of 5 per cent at the project level.

- ☆ Appropriate significance level i.e assigning probability to conclude that an observed change is a reflection of effort and did not occur by chance i.e at 95 per cent level

☆ Appropriate power *i.e.* the probability to conclude study has been able to detect a specified change *i.e.* at 80 per cent power.

☆ An assumed design effect of 1.5

Based on the above considerations the required sample size (n) for a variable of interest as a proportion for a given group was computed by;

$$n = D [Z_{1-\alpha} \sqrt{2P(1-P)} + Z_{1-\beta} \sqrt{P_1(1-P_1) + P_2(1-P_2)}]^2 / (P_2 - P_1)^2$$

where:

D= Design effect[5] (Assuming a design effect of 1.5)

P1= The estimated proportion at the time of the first survey;

P2= The proportion expected at the time of survey

Z1-α= The z-score corresponding to a significance level

Z1-β= The z-score corresponding to the power

In order to estimate change, we have taken base values of ever-married women who have ever experienced spousal violence from NFHS-III for each state. Based on the base value we computed the sample size require to detect a change of 10 per cent at state level. Based on the above formulae adequate sample size for state level assessment comes out to be in range of 500 to 600 at state level for women in the range of 15-50 years. The project proposes to take similar sample for men in the range of 15-50 years. The total sample size to estimate change is around 3800 *i.e.* 1900 men and women each in the range of 15-50 years. Around half of the sample of the project is taken from comparison villages.

In quantitative analysis, to match the project and non-project groups, propensity score matching (PSM) method used. The propensity score or the probability of participating in the program (being treated), is a function of the individual's observed characteristics $P(X) = Prob(D = 1 | X)$ Where, D indicates participation in project X is the set of observable characteristics. To measure the effect of the programme, we maintain the assumption of selection on observables *i.e.* assume that participation is independent of outcomes conditional on $X, E(Y | X, D = 1) = E(Y | X, D = 0)$ if there had not been a programme. This is false if there are unobserved outcomes affecting participation. Modified version of PSMATCH2 using STATA 10.0 module used to compute the average treatment effect across project and comparison area. The broad steps followed for the computation of the comparison estimates are described below:

☆ Creating a dichotomous variable for the two groups *i.e.* project and comparison area.

☆ Generating propensity score using probit estimation using pscore module of STATA to give each household a propensity score

---

5   In case of complex cluster sampling design, two key component of the design effect are intraclass correlation, and the cluster sample sizes. Thus, the design effect is calculated as follows:

Design effect = 1 + α?(n − 1) .

☆ Balancing the matched set of Household to ensure equality of means across block

☆ Computing Average treatment effect using Local Linear Regression matching and ensuring common support.

Disaggregation by Wealth Index or Standard of Living Index (SLI) used for conducting disaggregation across quantiles on a wealth index. Wealth index highlights the background characteristics about the economic status of households. The economic index is constructed using household asset data and housing characteristics. Each household asset is assigned a weight (factor score) generated through principal components analysis. Each household is then assigned a score for each asset, and the scores are summed for each household. The sample is then divided into three groups at equal intervals *i.e.* low, medium and high. Principal Component Analysis (PCA) is used as the main statistical tool along with reliability analysis which measures correlation amongst a set of indicators at different levels to shortlist, retain and categorize statistically relevant ones. The objective is to determine a single composite variable *i.e.* household score obtained after summing up asset score for each of the 33 household assets. Households are then ranked according to their total scores. Further, the wealth index constructed is segmented into different categories to help identify the more vulnerable sections within the target community. The range of scores is divided by calculating cut-off points such that the range of scores is divided into three groups at equal intervals.

Content analysis conducted for the qualitative data collected. The format for content analysis tried to capture experience and individual perception of women survivors of gender based violence facilitated through IPAP programme. Approximate 50 respondents across all states and 6 case studies are documented. End product of this study will also strive to establish whether the implementation has contributed to the project goal *i.e.* "Reduce the social acceptance of all violence against women and bring a positive change in the policy and programme environment that perpetuates its acceptance at an institutional and community level" and document effective use of MEL practices as per OXFAM India Country programming.

Key Findings from the Study

OXFAM's IPAP intervention addressed an issue which has been covertly and overtly present in our societies, supported ostensibly by elders and guardians. Women experiencing violence often are forced to remain silent and accept the violence as the formal justice institutions (police, courts) have failed to respond to women survivors and more towards vulnerable communities. According to National Family Health Survey (NFHS-3) conducted in 2005-06, "About 35 percent of women age 15-49 in India have experienced physical or sexual violence. Thirty-seven percent of ever-married women have experienced spousal physical or sexual violence and 16 percent have experienced spousal emotional violence."

The key findings of the evaluation are as follows:

Attitude Towards Violence Against Women and Knowledge of Laws Related to VAW

☆ Approximately 60 percent men and 61 percent of women from project areas against 55 percent men and 50 percent women believe that it is totally unjustified to beat wife. Comparison with the baseline also shows significant improvement. At the time of the baseline less than 1/3ʳᵈ of men and women believed this practice to be justified. Awareness index[6] of women and men on VAW was created using a set of indicators. About 8 percent men in the project area and 5 percent in comparison area lie in the high awareness category while 46 percent in the project and 43 percent in the comparison are in the moderate category. However, data on women across project and control areas does not show significant difference.

☆ The knowledge of the men and women on different provisions of law related to violence against women was also assessed. 37 percent of men had very low awareness about the laws related to VAW in the project area as compared to 39 percent in comparison areas. A comparatively high percentage of female respondents were not aware about the laws related to VAW both in the project (50 percent) and comparison areas (54 percent). The awareness of men in project areas about domestic violence act has increased from 24 percent in baseline to 31 percent in endline while the same for women has increased from 26 percent in baseline to 49 percent in endline.

Experience of Survivors

Findings from In-depth discussions with women survivors

☆ Physical abuse and verbal abuse are the most common forms of violence experienced by the survivors. Half of the survivors also report that they have experienced "Psychological Abuse/Undermining self-esteem/Embarrass/Offend in Public". Almost 78 percent of the respondents reported that they approached their "own family" to seek help from abuse or violence. More than half (53 percent) sought help from "NGO/Social Service Organization". 43 percent also mentioned that they approached the "Support Center". They also took help from Police, Lawyers, and Neighbours etc. The survivors were asked whether their violence stopped because of the intervening of the concerned persons they approached for help. Almost 78 percent of them reported "yes".

☆ Survivors have utilized different sources of help for registering their cases. Majority of the survivors received help either from their parents or support

---

6   The indicators that were used to calculate the awareness index are: understanding on physical abuse, verbal abuse, preventing the woman from obtaining employment or education, forcing the woman to hand over her income, psychological abuse, holding the woman captive at home, embarrass in public, slander or insult and forcing woman to have sex/rape.

centre counsellors and NGO staff members. Support centres definitely have played a significant role in registering of cases. The study findings suggest that police station-based counselling centres have led to better detection of cases of VAW.

Service Providers

Findings from In-depth discussions with police and WCD representatives,

☆ The major concern expressed by the police officers was to identify the extent of domestic violence within the localities in their jurisdiction though they believe that in the last few years reporting on domestic violence has shown an increase.

☆ Police officials across all the states have observed significant change in people's perception towards domestic violence. They suggest that women have opened up and cases of violence are getting reported to the police more frequently. But still there is a lot of hesitation in reporting sexual abuse as a form of domestic violence.

☆ Police officials across all the states show high level of awareness on the provisions under the PWDVA. In states like Andhra Pradesh and Gujarat, officials have had good exposure to training programs conducted by various civil society organisation and state government. These trainings have helped them to deal with identification and responding more effectively to the cases of domestic violence. Police personnel believe that their engagement in domestic violence cases have improved in the last few years.

☆ It is encouraging to state that WCD (Department of Women and Child Development) officials exhibit good level and depth of awareness on various legislations and provisions under law. While they show high level of awareness, their engagement as protection officer is not always their priority. They suggest that being a protection officer is just a small part of their overall responsibility as a WCD official. They also suggest that they lack clarity on their role as a protection officer. As their responsibility with the WCD involves a lot of travel, the issue of their limited reach for survivors adds up to this.

Community Groups

☆ IPAP programme was implemented on ground by a network of partner organizations/civil society organizations (CSOs) who executed the implementation plan on the ground. They were also engaged in developing program Pressure Groups/Vigilance Committees in the program communities. The groups are primarily engaged with the community for advocacy on domestic violence, identification and support to of victims and counselling.

☆ Approximately 60 percent of the community group members had participated in the trainings conducted by Oxfam during the last three years. Almost 60 percent of the members reported that they have more clarity on laws related to VAW now. 51 percent had mentioned that they had become more aware regarding which officials to contact for VAW cases. 43 percent reported that they now have more clarity on support services and facilities. 66 percent of group members suggest that they had helped women experiencing violence in their locality.

☆ To measure this, community group members were asked to react to statements which were related to husband hitting or beating wife under different situations. Three fourth of the community group members suggest the actions suggested in the statements are totally unjustified. Around 57 percent of the members had moderate level of awareness while 40 percent had high level of awareness on VAW.

Discussion: Effective Use of Mixed-Method Evaluation Research Design with Vignettes

The paper attempted to reviews and reflects on how mixed method quasi-experimental research design along with vignettes to capture untold experiences of human life under threat, at risk in particular among the women survivors of violence. Mixed methods (MM) evaluations seek to integrate social science disciplines with predominantly quantitative (QUANT) and predomi-nantly qualitative (QUAL) approaches to theory, data collection, data analysis and interpretation. The purpose is to strengthen the reliability of data, validity of the findings and recommendations, and to broaden and deepen our understanding of the processes through which program outcomes and impacts are achieved, and how these are affected by the context within which the program is implemented. While mixed methods are now widely used in program evaluation, and evaluation commissioning organizations frequently require their use, many evaluators do not utilize the full potential of the MM approach.

Vignettes have been used by researchers from a wide range of disciplines, yet very few methodological papers examine the use of the technique, particularly its application within qualitative research. Vignettes are short scenarios or stories in written or pictorial form which participants can comment upon. Whether researching in the 'qualitative' or 'quantitative' tradition, the central feature of this method is to explore participants' subjective belief systems. Finch (1987:105), for example, describes vignettes as "short stories about hypothetical characters in specified circumstances, to whose situations the interviewee is invited to respond". Hughes (1998:381) similarly defines them as "stories about individuals, situations and structures which can make reference to important points in the study of perceptions, beliefs and attitudes".

Vignettes may be used for three main purposes in social research: to allow actions in context to be explored; to clarify people's judgements; and to provide a less personal and therefore less threatening way of exploring sensitive topics. In qualitative research, vignettes enable participants to define the situation in their own terms. Multi-method approach - Vignettes have been widely used as a complementary technique alongside other data collection methods (see Hazel 1995; Hughes 1998). They can be employed either to enhance existing data or to generate data not tapped by other research methods (such as observation or interviews). With regard to the former, MacAuley (1996) sought to explore children's perceptions and experiences of long-term foster care, using vignettes, unfinished sentences, postal boxes, response cards, games and other visual stimuli to achieve an 'insider' position on children's perceptions and value systems. Wade (1999) employed vignettes following individual interviews in her study about children's perceptions of the family. She selected stories on topics that had not been covered in the interview or which would benefit from further exploration. Barter and Renold (1999), in their work on violence between young people in residential children's homes, also used vignettes in conjunction with semi-structured interviews. They routinely asked all participants to respond to a range of selected vignettes, regardless of whether they had disclosed a similar situation in the interview. In this way, a systematic comparison of individual responses to different behaviors could be generated.

The benefits of a mixed methods approach[7] encapsulated real time operational scale with many attributes. It narrates in understanding how local contextual factors help explain variations in program implementation and outcomes. Reconstructs baseline data for QUANT evaluations when it was not possible to conduct a baseline survey. Many evaluations are commissioned toward the end of the program and do not have very reliable information on the condi-tions of the project and comparison groups at the time the program began. This makes it difficult to determine whether observed differ-ences at the end of the project can be attributed to the effects of the program or whether these differences might be due, at least in part, to preexisting differences between the two groups. For example, women who apply for small business loans may come from families that are more supportive of women owning a small business than most families, or they may already have more business experience than women who do not apply for loans. If these preexisting differences are not identified, there is a risk of overestimating the effects of the loan program. It is often possible to use such QUAL techniques as in-depth interviews, key informant interviews or focus groups to obtain information of the characteristics of program beneficiaries and non-beneficiaries at the time the program began. This kind of information, which is often quite simple and economical to collect, can greatly enhance the validity of exclusively QUANT survey data. Strengthens the representativeness of in-depth QUAL studies (for example, by linking case study selection to the QUANT sampling frame) can make it easier to compare findings with QUANT survey data. Provides a good sense about validity and value of different kinds of QUANT and QUAL data. Promotes greater understanding of stake-holder perspectives on the nature of the intervention or how it is expected to achieve its objectives. This promotes a more participa-tory approach and greater alignment between stakeholders and evaluators.

In Multi-method approach Vignettes have been used in conjunction with other data collection methods (see Hazel 1995; Hughes 1998; Barter and Renold 2001). They can be used to generate data untapped by other methods (*e.g.* observation or

---

7   For a recent review of the benefits of mixed methods approaches see Adato (2012).

interview). In our evaluation study on understanding untold stories of violence and how women survivors did cope-up with real time situation at risk and addressed it, vignettes were used alongside semi-structured interviews.

There are possibilities identified as in when vignettes use showed good results. Those are like, Sensitive topics - We found vignettes to be especially useful in engaging young people and women survivors of violence to discuss potentially sensitive topics in a number of ways. For example, it was easier for those who did not want to discuss personal experiences to respond to other people's experiences. This was particularly true of some of the women survivors in our research expressed discomfort who found discussing their personal experiences openly. While in the contrast when respondents were given hypothetical story and situation to think over it, they developed intimacy by focusing on reading the vignette, seemed to create a comfortable distance between the researcher and participant and went some way to facilitate a nonthreatening environment. Compensating for lack of personal experience - Vignettes were employed to tap attitudes regardless of whether participants have had any direct experience of a situation. Having limited and unstructured knowledge of respondent's experience of violence, vignettes became an invaluable component of our methodology because they provided a focus for those who had no personal experience of violence in facing actual life threatening situation. Comparing disparate groups - The application of vignettes offered the opportunity to compare and contrast different groups' (*e.g.* treatment and control study group, heterogeneous respondents of male and female, staff and pupils, pupils from contrasting schools) interpretations of a 'uniform' situation, while at the same time enabled certain cultural factors such as age, gender and ethnicity. This was especially important in our understanding of the wider context in which different types of violence were perpetrated or experienced and permitted a more systematic development of benchmarks for understanding differences in interpretation.

Conclusion

This paper has reviewed and reflected of using mixed method quasi-experimental research design along with vignettes for a Gender Programme Evaluation at focused state of intervention. The essence of sharing this experience is to portray contrasting trend of using multi-method approaches which has established some or proportional empirical evidence to answer about behavioral causes of violence against women in any form exist. This has further discussed about how there can be different layers of mixed method – vignettes applied together in a quasi-experimental design. While adopting this technique will ultimately depend upon the aims and objectives of concerned research project and its funding for evaluation. It is experienced, if this method can be used as an integrated approach flexibly, vignettes can be particularly productive in a number of ways. They can: engage young people to discuss potentially sensitive research topics; enable an exploration of issues or incidents that participants may have no direct experience of; and systematically compare and contrast disparate groups' interpretations and belief systems, perceptions, attitudes on the unaddressed problem statement. It is concluded that empirical findings and quality of evaluation results can be documented if the research approach is justified with appropriate use of methodology of evaluation and linked to outcome of study.

Acknowledgement

The opinions expressed in this paper are those of the author(s) and do not necessarily reflect those of Oxfam. The paper is based on meta-review and endline evaluation conducted by Oxfam India during the programme phase out. The aim of the DfID supported INGO Partnerships Agreement Programme (IPAP) was to "improve the status of the poorest and most marginalized in India". Sincere thanks and appreciation goes to the partner organizations of the Department for International Development (DfID), UK Aid, supported International NGOs Partnership Agreement Programme (IPAP). This evaluation would not have been possible without their support and reflections. A special thanks to the communities who shared information and participated whole heartedly in the evaluation process.

References

Celine Sunny, 2003. "Domestic Violence against Women in Ernakulam District", Discussion Paper, Kerala Research Programme on Local Level Development, Centre for Development Studies, Thiruvananthapuram.

Change. Learning about the Human Rights of Women and Girls. New York: UNIFEM and the Center for Women's Global Leadership.

Dave A. and G. Slinky. 2000. Special Cell for Women and Children: A Research Study on Domestic Violence', in Domestic Violence in India 2: A Summary Report of Four Record Studies. Washington DC: International Centre for Research on Women and The Centre for Development and Population Activities.

Duvvury, N and M. B. Nayak. 2003. The Role of Men in Addressing Domestic Violence: Insights from India'. *Development*. 46(2): 45-50.

Hamberger, L.K; J.M.Lore; D. Bonge and D.F.Tolin. 1997. An Empirical Classification for Motivations for Domestic Violence. Violence Against Women, 3(4): 401-23.

Heise, L., Ellsberg, M and Gottemoeller, M. 1999. Ending Violence Against Women. Population.

Reports, Series L, No. 11. Baltimore, John Hopkins University School of Public Health, Population Information Program, December.

International Center for Research on Women (ICRW), 2000. "Domestic Violence in India: A Summary Report of a Multi-site Household Survey".

Bamberger, Michael (2012). Impact Evaluation Notes No., Introduction to mixed methods in impact evaluation.

International Centre for Research on Women (ICRW). 2000, "Domestic Violence in India", Washington, DC, May 2000.

International Clinical Epidemiologists Network (INCLEN). 2000. Domestic Violence in India 3: A Summary Report of a Multi-Site Household Survey. Washington, DC: International Centre for Development and Population Activities.

Oxfam India.2014. Promoting Violence-Free Lives for Women from Poor and Marginalized Communities in India: An Endline Evaluation report: India

Renold, Emma. 2002. Using vignettes in qualitative research. Cardiff University School of Social Sciences. *Building Research Capacity*. July 2002 Issue 3: ISSN 1475-4193

Faia, M. A. (1979). The Vagaries of the Vignette World: A Document on Alves and Rossi, American Journal of Sociology, 85, pp.951-54

Rahman, N. (1996). Caregivers' Sensitivity to Conflict: The Use of Vignette Methodology, Journal of Elder Abuse and Neglect, 8, pp.35-47.

Wilkinson, S. (1998). Focus Group Methodology: A Review, *International Journal of Social Research Methodology Theory and Practice*, 1 (3) pp.181-203.

Finch, J. (1987). The Vignette Technique in Survey Research, Sociology, 21, pp.105-14

Landsberg-Lewis I. (1998). Bringing Equality Home. Implementing the Convention on the Elimination of All Forms of Discrimination Against Women. New York: UNIFEM.

Ministry of HRD, 2000. "Platform for action violence against women- An assessment", Department of Women and Child Development, Government of India. Mertus J., Flowers N. and Dutt M (1999) Local Action, Global.

Mishra, J. 2000. Women and Human Rights. Chapter 5. Kalpaz Publications, New Delhi.

Mitra, N. 1999. Best Practices Among Responses to Domestic Violence in Maharashtra and Madhya Pradesh', in Domestic Violence in India 1: A Summary Report of Three Studies. Washington DC: International Centre for Research on Women and The Centre for Development and Population Activities.

Society for the Promotion of Ethical Clinical Trials – Scientific Review Board http://www.spect.in/independent-ethics-committee-in-india-spectsrb.

The DAC Principles for the Evaluation of Development Assistance, OECD (1991), Glossary of Terms Used in Evaluation, in 'Methods and Procedures in Aid Evaluation', OECD (1986), and the Glossary of Evaluation and Results Based Management (RBM) Terms, OECD (2000). [http://www.oecd.org/dac/evaluation/daccriteriaforevaluatingdevelopmentassistance].

Chapter 13

Evaluating Effectiveness of Elected Women Representatives of PRI in Monitoring Quality of Health Service Provision through Accountability Checklists

Aparajita Gogoi[1], Madhu Joshi[2] and Manju Katoch[3]

Center for Catalyzing Change (formerly CEDPA India)
New Delhi
E-mail: [1]agogoi@c3india.org, [2]mjoshi@c3india.org,
[3]mkatoch@c3india.org

ABSTRACT

Centre for Catalyzing Change (Formerly CEDPA India) with the support of David and Lucile Packard Foundation, initiated project *Pahel* (meaning 'initiative') in Bihar to strengthen the voice, participation, leadership and influence of Elected Women Representatives (EWR) by equipping them to address and monitor the quality of State run local Reproductive and Maternal Health (RMH) services.

The project is reaching out to 1200 EWRs with inputs not only on Panchayati Raj Institution (PRI) structures and health delivery systems but also on gender and patriarchy. The EWRs are using accountability checklists for monitoring services, and are using the findings to demand accountability at appropriate forums such as *Panchayats* and with health authorities.

The paper shares the experience of tracking of changes among EWRs in terms of their participation and initiative in PRI processes and monitoring health services due to intervention in the project area and corresponding shifts in their individual and collective self-efficacy and leadership skills using Most Significant Change (MSC) methodology. It also demonstrates how the intervention has capacitated EWRs to evolve both as health advocates and leaders in their constituencies and have been able to effect a visible change in the delivery of RMH and other health services over a period of two years.

Keywords: *Elected women representatives, Accountability checklists, Most significant change, Self-efficacy, Agency, Leadership skills*

Background

The objective of this paper is to demonstrate the use of qualitative and utilization focused tools to evaluate the impact/change in both the beneficiaries (Elected Women Representatives) and in the State run Reproductive and Maternal Health (RMH) services in six blocks of three districts in the state of Bihar.

The intervention, *Pahel:* Towards Empowering Women, initiated by Centre for Catalyzing Change (Formerly CEDPA India), in the year 2007 and supported by the Packard Foundation aims to strengthen the leadership skills of Elected Women Representatives by building their capacities and supporting them in monitoring and advocating for the quality of services being delivered by State run health facilities, to bring about social change in areas that affect women, particularly their reproductive health, thereby fulfilling a dual objective of women's political empowerment and State accountability for health services.

Ensuring Accountability for Health Services Under the National Rural Health Mission

The National Health Mission, earlier known as the National Rural Health Mission (NRHM) is an initiative by Government of India to ensure effective health care to communities living in underserved rural areas through a range of interventions targeting individuals, households, community and most critically the health system. The thrust of the mission is on establishing a fully functional, community owned, decentralized health delivery system.

Community based monitoring (CBM), now renamed as Community Action on Health, are recommended and mandated under NRHM as a strategy to ensure that basic health services reach the underserved segments of populations and transparency and accountability for these services is ensured at all levels. Within this framework, *Panchayati Raj* Institutions (PRI) members have a defined responsibility in monitoring and supervising health functionaries and enhancing the use, accessibility and quality of available services. However, the experience of CBM has been mixed – an assessment of the first phase (initiated in 2007) across nine states indicates that a lot needs to be done in terms of capacity building, convergence of departments, and recommends that the process still needs significant nurturing and direction from Ministry of Health and Family Welfare (MoHFW). Additionally, there are recommendations that the process and tools should be simplified to enable its use by the community. Other challenges have been identified, including requirements of investments in intermediaries like Non-Government Organisations (NGOs) to hand hold communities, loss of time from data gathering to feeding back findings to effect change, and lack of expertise in the community to lead the process. There is a need for technical and financial support to ensure that the process continues to be implemented in the first phase states as well as initiated in other states. It further identifies engagement of PRIs as the weakest links in the process.

Project Pahel Strategies

Pahel is being implemented in the state of Bihar which poses a significant challenge. Despite, high political will that led to several inclusive, pro-people and effective governance policies, schemes, and reforms that have led to improvement in the several development indicators, Bihar with its population of 103.8 million (Census 2011), continues to be one of the poorest states in the country including a high Maternal Mortality Ratio (261 per 100,000 live births – SRS 2012) as compared to the national average of 178 and a continued high Total Fertility Rate (3.7) that further impacts on the MMR in the state.

Bihar is also the first state to formally reserve 50 per cent of seats for women in local governance. The state held its 3rd local body elections in 2011, with 50 per cent of seats reserved for women. There are currently more than 120,000 EWRs in the state at the three levels of local governance (district – *Zilla Parishad*, block- *Panchayat Samiti* and village – *Gram Panchayat*).

A baseline conducted as a part of the situational analysis revealed that 90 per cent of the 1200 EWRs, the project engages with, had not received any training on governance during their current tenure. 84 per cent of these EWRs were first time elects and had not been in power long enough to have had self learnt through exposure to the role of an elected representative. Poor literacy level among this cohort (40 per cent had never been to school, and only 29 per cent had studied between 1st-5th grade) further added to the challenges of a meaningful collaboration.

Capacity Building of EWRs

The objective was to build the capacities of EWRs to identify systemic gaps that hinder delivery of quality RMH services and then use the findings to raise the issues

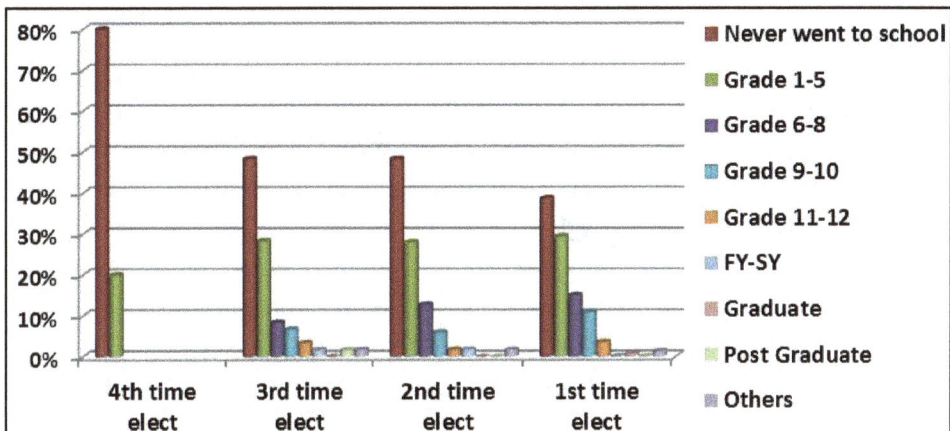

Figure 13.1: Profile of EWRs.

at appropriate forums such as *Panchayat* level meetings including meetings of *Panchayat* Standing Committees, meetings with health service providers, and with health authorities.

A three-day training on PRI structures/processes, gender and patriarchy and public health delivery system, specifically, Reproductive and Maternal Health (RMH) was conducted to enable the EWRs to understand the context.

The IPHS and NRHM guidelines and facility assessment survey formats (checklists) were fine-tuned to develop the checklist to be used under the intervention to evaluate adherence to government-stipulated guidelines for the four levels facilities— the monthly Village Health Sanitation and Nutrition Day (VHSND), the Health Sub-Center (HSC), the Primary Health Center (PHC), and the District Hospital (DH). The EWRs were also trained in the use of these checklists.

Use of Checklists to Monitor Health Services

Following their training the EWRs visited the health facilities within their jurisdiction, and administered the checklists to assess facilities on 6 key indicators relating to RMH services - infrastructure, personnel, community participation, availability of equipment, drugs and other supplies, service provision, and quality of logistical arrangements. The checklists were administered twice to track improvements in outreach as well as access, quality and availability of health services over time.

Given the low literacy levels, they required hand hold support from project Field animators who continuously guided them in their role as community representatives.

The checklist findings were developed into simple and specific advocacy asks, listed in the order of priority, and aggregated at HSC, PHC and DH level to help the EWRs gain a better understanding of the issues and gaps in health services.

Informal bi-monthly meetings called *Mahila Sabhas* were organized to mentor and collectivise the EWRs so that they could understand and share the issues around service delivery and their role as elected leaders of their communities, and also build solidarity as a peer group. These were strategic forums that helped them participate more effectively in *Panchayat* meetings, interface with government officials/health service providers and mobilise community women to access SRH services.

Block and district level convergence meetings had been designed under the project as platforms for advocacy, where EWRs' shared the advocacy asks with officials from different departments (Health, Social Welfare and Public Health Engineering Department/PHED).

Results and Measuring Change

The project continued to generate interest amongst EWRs, the tracking data showed that 127 EWRs who had not attended the trainings in year 1 joined the process during year 2 and additionally, 43 EWRs joined the process during year 3. In addition to health issues, the women began to initiate action in other issues like education, infrastructure and regular conduct of PRI meetings.

The monitoring and evaluation framework for the project included consisted of a quantitative design that tracked individual women for their participation in PRI processes such as meetings, interfacing and raising issues relating to SRH services. At the baseline the women were tested along the same parameters. A mid- term qualitative evaluation was conducted later to measure the quality of change among the EWRs – Most Significant Change (MSC) technique was used for this purpose. The MSC gathered stories from 30 EWRs from all three levels of PRIs and interviews with people from their circle of influence including husbands, health workers and male colleagues from Panchayats. Detailed story guides were prepared to capture changes in their lives since the inception of the intervention – lifestyle, journey as an elected political leader, attitude towards social issues like girls education, dowry, their motivation to work and strategies to overcome the barriers of patriarchy and illiteracy. The testimonies of In-Depth Interviews with others, validated the EWRs sense of accomplishment and growth as a consequence of Pahel.

Impact on the EWRs

Participation in the Process

79.3 per cent of the EWRs administered the checklist in Year 3 as compared to 59 per cent during year 1. Over a period of time with the raised confidence level the EWRs were able to observe, analyze and then record the information.

Participation in Meetings

The changes were reported by their male colleagues, "If there was no reservation women would have been invisible in politics, the *Pahel* women (the EWRs) actively participate in meetings. The EWRs have raised issues on FP/RH and conducted regular follow-ups during the *Gram Sabha* and urged the *Mukhiya* (*Gram Sabha* Chairperson) to place the gaps identified at the level of the local HSC during the *Panchayat Samiti* bi-monthly meeting to ensure action by the Medical Officer In-Charge. The two annual district level convergence meetings and one round of block level convergence meetings have been organized and have seen the active participation of EWRs from all levels. In one block this meeting was followed up with an interface with the local Member of Legislative Assembly (MLA) where women shared their findings and barriers to implementation.

> *"Earlier my life was limited to household work like cooking, feeding children and taking care of them, etc. I too was identified by my husband and father-in-law's name like the other women. Now I have come out of my routine housewife role and I go to block and speak to BDO or attend meetings and talk in public or speak to panchayat sachiv. I have got my own identity, people identify me by my name. This is a big change and I feel proud of it."*
>
> Ward member, Sitamarhi.

Self-worth and Agency

Through a qualitative mid-term evaluation using MSC, the project implementers attempted to gauge the motivation and extent of change in these women in the realm of their personal, professional and social life.

An EWR's husband shared "she (his wife, the EWR) has begun voicing her opinion on everything and is very well informed." The process of engaging in activities like monitoring health services, participating in *Mahila Sabhas* and speaking at PRI meetings have instilled a new sense of self-worth and strengthened the agency of these erstwhile dormant political leaders, who have now taken this further by addressing other development issues in their villages and ensuring transparency in governance.

> *"I can talk to Government officials and village level functionaries without hesitation. For instance, I can tell them that in my ward there are no street lights, no drainage facility, etc. I also visit the hospital if I get any complaint, for instance, if someone from my ward went to hospital and s/he did not get treatment or medicine. During my visit i talk to hospital staff and at times I tell them that they are public servants and can't leave their work and go whenever they feel."*
>
> Ward member, Aurangabad.

Impact on Health Service Delivery

Monitoring data under the project shows that the EWRs initiatives have yielded results in the realm of health service delivery– improving maternal health services by using untied funds to facilitate purchase of weighing machines, Sphygmomanometer (Blood Pressure measuring machine), examination tables, etc.

> *"I attend meetings conducted by ANM and talk to them and assess availability of medical supplies and take up the matter with appropriate authorities. For instance, ANM did not have equipment like weighing scale and BP measurement so I spoke to the medical officer about it. I was told that untied fund was available (Rs. 10,000/-) for minor purchases and repairs. I asked the ANM why she had not availed it. Similarly, if there is an expectant mother in my ward I call 102 for ambulance for transporting her to health facility for delivery if I am not around I tell them to call 102 whenever required."*
>
> Ward member, Sitamarhi.

Over time, the interface between the EWRs and health workers has become more frequent, their relationship has improved and a level of mutual trust has evolved. They have together found solutions – such as installing curtains for privacy during Ante Natal Checkup (ANC), and using untied funds for purchase of basic equipment. The health workers in turn have requested for the EWRs assistance for mobilization and awareness generation on FP/RH issues.

> *"Our EWR is an educated woman she speaks well and people listen to her. She raises issues related to women and girls and also tells me that she has discussed about the issue with mukhiya and it will be solved soon. I also discuss with her the issue of cleanliness in the community, illnesses and medical supplies and seek her help whenever required."*
>
> – Auxiliary Nurse Midwife (ANM), Aurangabad.

Visible improvement is seen in participation during VHSNDs at the level of community, EWRs and health workers. VHSNDs are being held regularly and privacy and quality of service delivery has improved during VHSNDs.

The EWRs have also been able to effect some improvements in service delivery at the 109 HSCs through their regular visits and interaction with the health workers. To report a few, availability of functional weighing machines and BP apparatus has improved by 37 per cent each.

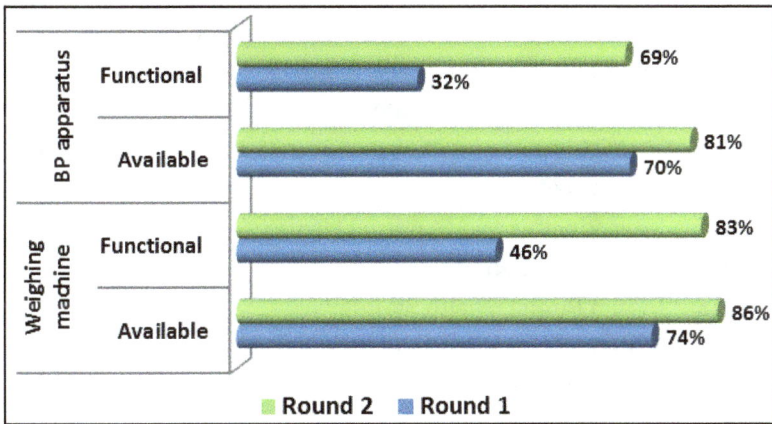

Figure 13.2: Availability of Functional Equipment.

Conclusion

The experience of measuring the results of project *Pahel* in terms of both emergence of EWRs as effective leaders in their constituencies and their consequent impact on ensuring State accountability for health services highlights the need for developing both quantitative and qualitative tools, with the latter using a feminist, rights based approach.

The qualitative assessment using MSC has been able to capture the levels and motivations for change in the EWRs who have emerged as effective leaders. It very clearly shows that women's leadership skills for governance can be built through continuous and consistent support and mentoring. Processes of empowerment instilled a sense of self-worth and strengthen the agency and also help build the capacities of women elected into the political system. The women themselves are quick to learn and act, both individually and collectively, in their capacity and are able to question structures and power relations within the family and outside the family. The new found confidence and information about their powers enable them to further address other development issues in their villages and ensure transparency in governance.

Alongside, hybrid forms of social accountability, with an emphasis on community mobilisation, hold the potential to empower women to assert their entitlements to health. It also opens possibilities for better service-user interfaces and opportunities for co-operation between the two. The *Pahel* experience reinforces the thesis advanced by Quinn-Patton that "the intended users are more likely to use evaluations if they understand and feel ownership of the evaluation process and findings; they are more likely to understand and feel ownership if they have been actively involved; and by actively involving primary intended users, the evaluator is training users in use, preparing the groundwork for use, and reinforcing the intended utility of the evaluation every step along the way (Patton, 2001).

Abbreviations

ANM:	Auxiliary Nurse Midwife
CBM:	Community Based Monitoring
CEDPA:	Centre for Development and Population Activities
CHC:	Community Health Center
DH:	District Hospital
EWR:	Elected Women Representative
RMH:	Reproductive and Maternal Health
HSC:	Health Sub-Center
IPHS:	Indian Public Health Standards
MMR:	Maternal Mortality Ratio
MoHFW:	Ministry of Health and Family Welfare
NGO:	Non-Government Organisation
NRHM:	National Rural Health Mission
PHC:	Primary Health Center
PRI:	*Panchayati Raj* Institutions
VHSND:	Village Health Sanitation and Nutrition Day

References

1. Centre for Health and Social Justice. 2010. *Reviving Hopes, Realising Rights: A Report of the First Phase of Community Monitoring under NRHM.*

2. Murthy, R.K. 2008. *Strengthening accountability to citizens on gender and health.* Global Public Health, 3, (suppl 1).

3. Patton, M.Q. 2001. *Qualitative Research and Evaluation Methods (2nd Edition).* Thousand oaks, CA: Sage Publications.

4. Posani, B. Aiyar, Y. 2009. *Accountability Initiative.* AI Working Paper No. 2.

5. Susan A. Papp SA, Gogoi A and Campbell C. 2012. *Improving maternal health through social accountability: A case study from Orissa, India.* Global Public Health: An International Journal for Research, Policy and Practice.

6. World Bank. 2004. *World Development Report 2004: Making services work for poor people* [Internet]. World Bank. Available from: http://web.worldbank.org/WBSITE/EXTERNAL/EXTDEC/EXTRESEARCH/EXTWDRS/0,contentMDK:23083461~pagePK: 478093~piPK:477627~theSitePK: 477624,00.html

Chapter 14

Experience of Evaluating Gender Integration in Livelihoods Improvement for Economic Security (LIFE) Project, Nilgiris District, Tamil Nadu

D. Yeswanth[1] and Sashi Kumar[2]

Microfinance Coordinator,
Care India Solutions for Sustainable Development (CISSD)
No. 36, 2nd Main Road, Kalaimagal Nagar
Ekkaduthangal, Chennai – 600 032, T.N.
E-mail: sashik@careindia.org

ABSTRACT

Gender Integration framework of CARE is to strengthen and promote gender equality by ensuring that we systematically assess the differential impacts on women and men through our intervention. Women empowerment framework and Strategic Programing Framework of CARE India was used as base material to develop GED tool. With support from Teavana, *Livelihoods Improvement for Economic Security Project* was implemented in two blocks of Nilgiris District, India. LIFE project goal was to improve productivity, incomes of 1000 tribal households involved in tea cultivation. Applying gender integration lens to the work of the LIFE project helps look at the potential impact of the program, policies and decisions may have on men and women. By introducing gender integration framework within its projects, CARE INDIA proposes to take a systematic approach to ensure that gender and intersectional issues are considered in all aspects of its work.

Five Focus Group Discussion and 30 individual surveys was used to collect information from the tribal community. Tools used include Seasonality exercise, Network analysis, Wage analysis and Life cycle analysis. Collected data was analysed based on 23 identified key dimensions of social change covering all three aspects (Agency, Structure and Relations) of GED tool. Observation reveal that as the project was more focused towards economic improvement, livelihood security of tribals and operate within specific deliverables - gender integration became a challenge to achieve in all the identified areas. It is suggested that application of gender perspective starting from the design of the project could show better results.

Keywords: *Gender integration, Livelihood project, Tribal community, GED framework (Agency, structure and relation).*

Background

CARE has been working in India for over 60 years, focusing on ending poverty and social injustice. We do this through well planned and comprehensive programmes in health, education, livelihoods and disaster preparedness and response. Our overall goal is the empowerment of women and girls from poor and marginalised communities leading to improvement in their lives and livelihoods. We are part of the CARE International Confederation working in 87 countries for a world where all people live with dignity and security.

Gender Integration and Women Empowerment

The Gender Integration framework is to strengthen and promote our commitment to gender equality by ensuring that we systematically assess the differential impacts on women and men through our intervention. It also ensures that we consider any adverse impacts produced by other intersecting grounds. CARE INDIA is committed to uncover and transform the political, social and economic relationships at the heart of poverty with the aim to improve the health and well-being of women and girls. At CARE INDIA, we view women's empowerment through the lens of poor women's struggles to achieve their full and equal human rights. In these struggles, women strive to balance practical, daily, individual achievements with strategic, collective, long-term work to challenge biased social rules and institutions. Therefore, CARE INDIA defines women's empowerment as the sum total of changes needed for a woman to realize her full human rights – the interplay of changes lie in:

☆ **Agency**: her own aspirations and capabilities,

☆ **Structure**: the environment that surrounds and conditions her choices,

☆ **Relations**: the power relations through which she negotiates her path.

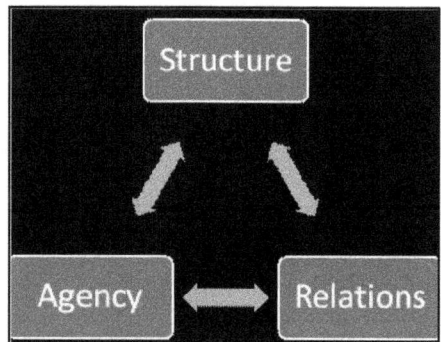

CARE's global framework links women's own definitions and priorities for empowerment to 23 key dimensions of social change which have been shown to be widely relevant to women's empowerment across many studies and contexts and is summarized below.

Agency	Structure	Relations
1. Self-Image; self-esteem	11. Marriage and kinship rules, norms, processes	19. Consciousness of self and others as inter-dependent
2. Legal and rights awareness	12. Laws and practices of citizenship.	20. Negotiation, accommodation habits
3. Information and skills	13. Information and access to services	21. Alliance and coalition habits
4. Education	14. Access to justice, enforceability of rights	22. Pursuit, acceptance of accountability
5. Employment/control of own labour	15. Market accessibility	23. New social forms: altered relationships and behaviours
6. Mobility in public space	16. Political representation	
7. Decision influence in household	17. State budgeting practices	
8. Group membership and activism	18. Civil society representation	
9. Material assets owned		
10. Body health and bodily integrity		

Gender Integration in LIFE Project

About Livelihoods Improvement for Economic Security (LIFE) Project

With support from Teavana, in 2010 CARE INDIA initiated the project in the blocks of Kothagiri and Gudalur of Nilgiris District in Tamilnadu, India. Goal of LIFE project is to improve the productivity and incomes of 1000 tribal households involved in tea cultivation. While the entire family benefit from the project, CARE particularly focuses on working with women not only because they are disproportionately affected by the poverty, but also because they can serve as powerful agents of change in their families and communities when equipped with the right tools and resources.

Specific Objective of the Project was to

☆ Organize 1000 tribal small tea growers into groups through which individuals can gain access to services and inputs.

☆ Increase the productivity of 1000 tribal tea farmers through adoption of better cultivation practices.

☆ Diversify the livelihoods of tribal household mainly to reduce dependence on a single crop (tea) and thus encourage more balanced nutrition uptake by the community.

☆ Pilot organic tea cultivation with tribal small tea growers mainly to explore possibilities for expansion and link to niche markets.

Applying a gender integration lens to the work of the LIFE project, Nilgiris helps look at the potential impact of the program, policies and decisions may have on men

and women. Does it recognizes sex and gender, as well as factors affecting women in tribal community, adolescent girls and children's? By introducing gender integration framework within its projects, CARE INDIA proposes to take a systematic approach to ensure that gender and intersectional issues are considered in all aspects of its work.

Study Methodology

Women empowerment framework and Strategic Programing Framework (SPF) of CARE INDIA was used as base material to develop the GED tool. CARE INDIA team discussed the framework in detail and worked together to segregate the indicators under three major themes namely, ***Agency, Structure and Relationship***.

Indicators developed include aspects like household chores management, access to productive resources, access to financial services and social security schemes, negotiation/mechanisms for benefit sharing, reproductive role related, production related, working condition, participation in decision making at household and community level and overall public participation. Feedback was received from program implementation team, field staff and partner agency staff to finalise the indicators classified under each theme.

Gender analysis process has been classified into 7 steps which are listed below

1. **Consultations:** Internal consultation process was carried out to commence the study. During consultation the study process was designed and possibilities of adequate representation of women and men among respondents were explored.

2. **Defining the Issues:** Second level process include designing strategy to approach the community based on the identified issues *E.g.* Are both women's and men's experiences reflected in the way issues are identified?

3. **Gathering Information:** What types of gender-specific data are available? How will the proposed study will help address different experiences of women and men?

4. **Conducting Study:** What is the research methodology? Are additional areas of research needed to obtain enough information relevant to both women and men?

5. **Developing and Analysing Options:** How will each option have a different effect on women's or men's social and/or economic situation? How will innovative solutions be developed to address the gender issues identified?

6. **Data analysis:** How will gender equality concerns be incorporated into the evaluation criteria? How can this be demonstrated? What indicators is used to measure the effects of the policy or program on women and men?

7. **Communication:** How will the communications strategy ensure that information is accessible to both women and men?

Tools Used During Exercise

To collect information from the community two type of methodology was used

1. Focus Group Discussion (FGD) and
2. Individual interview method was used.

FGD was organised in five villages and in each village six individual household interview was conducted to collect the information. In total, five FGD and 30 individual questionnaire was done by the team to understand the gender integration in the LIFE project.

To conduct FGD the following tools were adopted to collect the information. It includes

☆ **Seasonality exercise** – Where the income, expenditure and savings, borrowing pattern was collected month-wise. In the absence of knowledge about the month (English/Tamil calendar) it was decided to collect the above information from the community based on festival celebration time.

☆ **Network analysis** – Different stakeholders involved in the production and marketing process of tea production and also access to different services was collected using Venn diagram mainly to understand the network with which the tribal community interact.

☆ **Wage analysis** – Tribal community rely on different occupation to earn income. Wage analysis focused on the difference in wage pattern between men and women, activity wise allotment of work among the household, facility available to men and women in their work.

☆ **Life cycle analysis** – To collect detailed information, the life cycle phases (Birth, youth, marriage and death) was used as a reference point. Above method helped the community to reflect on the questions in a more detailed way and help recognize the importance with reference to the GED process.

Gender Integration LIFE Project

Progress in Structure

There is gradual behavioural change among tribal women especially in the project intervention areas. In LIFE project, women members are mobilised under village level groups (VLG) and are encouraged to represent in the group meetings. LIFE project has created leadership opportunity for representation, participation and access to formal financial institutions (bank) by encouraging women to approach and do transactions with the bank along with a male companion. Visible changes are observed in-terms of women representation and their attendance in public meetings after the LIFE intervention, whereas it was almost nil/absent when the initiative started.

Women participation in voicing their views and taking active part in community level decision making is low at present but for the tribal community their mere attendance in meeting is considered as a big transformation as a whole. Timeline analysis reveal that male members and elders of the community has started giving space in common meets for women and recognise their importance as a member in Village Level Group. Recent time's women participate more freely both within and outside the society but there is a need to work further to ensure their voice/opinion are counted at larger level.

Way forward women decision making is largely restricted to household level however emphasis should be insisted for larger community level participation with focus on equality approach. A continued effort is required further to improve participation of all the women (within the VLG). At present only a few - mainly office bearers present their view, take up responsibilities and not all of them. Increased women participation and providing opportunity for leadership would bring balanced structure resulting in improved relations.

Community Strengthening

Strengthening community led institutions indirectly encourage public participation of men and women, that help create identity among external environment involving various stakeholders. The agency level intervention help bring change within the community but it is a time consuming process mainly due to the constraint of bringing about 'behavioural change' among the various stakeholders. At present there is increased collectiveness among the tribal, for instance after the LIFE intervention they actively come together to represent their requirements with external stakeholders - meeting government officials for arranging water supply, sanitation facility etc.

There has been lot of investment under the project in creating a structure but the community strengthening process has a long way to go.

   ☆ It is observed that the community as a whole has very weak negotiation or bargaining power and therefore the collectivizing the community has resulted in very limited impact for the population.

   ☆ Further the external environment stereotype thinking about tribal has to change.

Increasing women population representation in such village level organisation would further strengthen the process. At present in LIFE project only about 40 per cent have women representation; since men own and had more opportunity to get entitlements men were brought under the project. At structure level continuous support, regular community strengthening and capacity building would enhance capacity of individuals. Relations could be improved through better awareness and understanding on basic rights and their entitlements.

Capacity Building at Individual Level

At agency level there needs a significant intervention on engaging girl and boy child who have different need and expectation. Similarly youth and adolescent populations' vulnerability is completely different like child marriage, child labour and exploitations in workplace (in terms of safety, working hours, middlemen interference). Intervention with particular focus on adolescent in terms of identifying their skill requirements and gaps through capacity building will help increase their livelihood opportunities and would improve their economic conditions. Indirectly this would also help in restricting migratory job attitude and reduce exposure to vulnerability. LIFE project has advocated bringing land title, house deeds in the women household members' name. Above resulted in behaviours changes within and outside the households for instance, how women are perceived by the society as a whole in the changing context.

Detailed assessment based on 23 identified key dimensions of social change covering all the three aspects (Agency, Structure and Relations) for LIFE project was done by the team.

Challenges Faced

☆ The team faced challenge in avoiding overlap between questions that are classified within the three heads of agency, structure and relations. Detailed discussion/clarification was made to build a common understanding within the team on the questions.

☆ During FGD, participation was less among general women population across all the villages. Women involved in the LIFE project was forthcoming in discussing the issues, whereas others were not ready to share their views and opinion.

☆ In the presence of men and elderly members of the village during the FGD, women were not ready to share or opine on many sensitive issues. (Child birth, family planning, household management).

☆ In many instance their participation often resulted in conflict with male members and their opinion was shunted with counter allegations by male.

☆ Aspects with reference to how they negotiate, strategies adopted to manage their household and benefit sharing mechanisms was very hard to be capture during FGD and requires further fine-tuning.

☆ In network analysis, the production related aspects was explored in-depth but focus on social aspects was minimum and requires in-depth understanding of the community.

While doing seasonality analysis team found difficulty in making participants realise about different months in a calendar year. The team addressed the issue by using festival time to correlate with different months

Key Learning for Future

To summarise, it was observed that LIFE project has integrated the gender as well as cross cutting themes within its approach and activities. Analysis using the framework highlight that integration of gender approach have positively influenced both at individual project participant and community level but is slow and more time is required to understand the overall change. Observation reveal that as the project was more focused towards economic improvement and livelihood security of tribals that operate within few specific deliverables the integration became a challenge. Further it highlights that application of gender perspective from the design of the project could show better results.

The study has provided a wide range of learning listed are the highlighted one which has been classified based on different stages of the study

Stage I – Pre Execution

☆ Before execution of GED tool in field, the field team required to be familiarised with the tool and build a common understanding on the exercise. Separate discussion is essential to ensure all the questions designed are understood in the same way by all team members.

Stage II – Execution

☆ The team should comprise of individuals with adequate knowledge on customs, norms and values of the community to ensure the diversity factor.

Stage III – Post Execution

☆ As the execution of the study was focusing on two methodology FGD and individual interview, it is important to focus on common scale in ranking mainly to avoid bias in interpreting the responses.

☆ Triangulation of the collected surveyed data with appropriate quality check will provide better understanding on the quantitative findings.

Acknowledgement

Author gratefully acknowledge the support and feedback received from Mr. Devaprakash, Regional Program Director, Chennai, Project team staff (Mr Chandrasegaran and Ms. Arputha Sheela), field team members and all household members of LIFE project.

EVALUATIONS IN FRAGILE CONDITIONS (HUMANITARIAN CONDITIONS/HARD TO REACH)

Chapter 15

Monitoring and Evaluating Humanitarian Aid: Are Aid Information Management Systems an Adequate Tool?

Neha Kumra

NILERD (formerly IAMR), Delhi
E-mail: neha.kumra@gmail.com

ABSTRACT

This paper provides an overview of Aid Information Management Systems (AIMS) in the context of humanitarian crisis and assesses the adequacy of AIMS with regards to implementation monitoring and performance evaluation of humanitarian resource allocation. Humanitarian evaluations need to assess the coverage of population groups facing life-threatening suffering, wherever they are marginalized geographically, socio-economically, or by virtue of their social location (Bamberger and Segone[1]). However, AIMS are not real time and data generated may be inaccurate in humanitarian situations. AIMS may be characterized by a focus on the technical aspects of the project to the detriment of the change management aspect of the project. AIMS are essentially 'social systems' including technical and non-technical aspects and dependent on organisational and environmental contexts (Heeks 2006). In the context of humanitarian situations, it is important to align interventions with human rights conside-rations. There is evidence that AIMS may be characterized by weakness in analysing, understanding and managing evolutions due to the limited presence of teams in the field and limited travel; burdensome requirements weighing down Country Coordination Mechanisms (CCMs), who already have restricted capacities; limited ability to adapt ongoing grants to changing situations; highly demanding coordination and financial management needs with respect to the limited capacities of implementing structures and of the overall environment (Solthis 2014). Nevertheless, a comprehensive information management system which provides information on both humanitarian and development assistance in a post-crisis scenario is desirable for assessing sustainability of humanitarian interventions.

---

1 http://mymande.org/sites/default/files/EWP5_Equity_focused_evaluations.pdf, accessed in December 2014.

Introduction

"In fast evolving development contexts or in emerging, ongoing or post-crisis environments, the development plan needs to be dynamic and revised and improved over time. Whenever development plans are updated during implementation, it is necessary to document the rationale for such changes. Effective monitoring and evaluation is important as it provides evidence to base such changes through informed management decisions" (UNDP, 2010)

The poorest and most marginalized people are disproportionately affected by disasters as their ability to cope is limited by poor living conditions, inadequate infrastructure, a lack of income diversification and limited access to basic services, especially education and information. An equal participation of all segments of society in disaster risk reduction decisions can help address the root causes of disasters and this in turn can address people's underlying vulnerabilities, increase their capacities to cope with the effects of natural hazards, and facilitate empowerment processes (UNDP 2010).

Humanitarian crises often create a category of those worst-off in society, that is, those most affected by the crisis. A humanitarian evaluation, therefore, needs to be equity-focused (Bamberger and Segone[2]) and needs to assess the relevance, effectiveness, efficiency, impact and sustainability – as well as the coverage, connectedness and coherence – of poli-cies, programmes and projects concerned with achieving equitable development results.

Humanitarian evaluations also need to assess measures to provide protection to the affected population. In many emergency situations, the population under threat needs to be protected from murder and harassment, as well as from discrimination that can lead to exclusion from basic services. The evaluation also needs to assess what measures have been taken to mitigate potential negative consequences of the humanitarian programme (Hallam, 1998).

Furthermore, humanitarian situations require real time data for monitoring and evaluating. Real time evaluations are often undertaken at an early stage of an initiative to provide managers with timely feedback in order to make an immediate difference to the initiative (UNDP 2010). They provide implementing staff with the opportunity to analyse whether the initial response or recovery is appropriate in terms of desired results and process. They can also be used in crisis settings where there may be constraints, such as absence of baseline data, limited data collection efforts due to a rapid turnover of staff members (for example, lack of institutional memory) and difficulty conducting interviews and surveys due to security issues, in conducting lengthier evaluations. It is important that equity-focused humanitarian evaluations emphasize accountability to the affected populations including the most vulnerable – children and women.

---

2 http://mymande.org/sites/default/files/EWP5_Equity_focused_evaluations.pdf, accessed in December 2014.

Aid Information Management System (AIMS)

AIMS are software applications that record and process information about development activities and related aid flows in a given country, in order to assist the recipient country in managing the aid it receives. The primary users of AIMS are government staff in central ministries who are responsible for coordinating and reporting aid, and for planning aid in relation to the national budget and national plans. The secondary users are those responsible for reporting the data required, including development partners.

AIMS fulfil two purposes at the same time: they strengthen a government's capacity to plan, implement, monitor and evaluate the use of public resources; and they enable aid coordination, information-sharing, and domestic and mutual accountability (Gabriel *et al.*, 2006). These two purposes potentially can be in conflict with each other in a post-conflict or fragile setting. While the international community urgently needs information about ongoing financial flows, complaints about the lack of information for planning and decision-making are common in emergency and humanitarian situations. Simultaneously, government capacity to adequately manage the use of public resources is usually severely diminished immediately after a crisis.

The following section lists the pre-requisites for a well-functioning AIMS (Gabriel *et al.*, 2006).

Pre-Requisites for a Well-Functioning AIMS

A commitment to mutual accountability and transparency is crucial for a well-functioning AIMS. Donors need to commit to reliable, timely and transparent data on aid flows, and recipient Governments need to ensure that there is transparency in data tracking and accountability to national constituencies (Parliaments and civil society) and donors for resource allocation. This requires that national development strategies are clearly defined through dialogue among national stakeholders and donors. The operational implications of increased ownership need to be acknowledged. In addition, the success of an AIMS will depend on clarity on its purpose, its best placement in the government institutional system, capacity required to make it work, donor 'buy in', realistic expectations of the tool, improvements in the budget and aid management processes, strengthening of IT infrastructure, and enhancing outreach and analysis capabilities.

Second, a well functioning AIMS will require appropriate institutional arrangements and human capacity. Most governments establish an Aid Management Unit within a core Ministry, usually the Ministry of Finance or Ministry of Planning. It is important, however, to reinforce existing departments such as the budget department before creating new units. Moreover, institutional location of the team responsible for the aid information management system is critical in determining its effectiveness. Clear inter-ministerial and inter-departmental responsibilities and reporting lines are essential. In many cases, the capacities of the department responsible for aid coordination are limited and often already overstretched. Analytical, communication and outreach, as well as negotiation capacities often need to be increased in order to use the system effectively and additional human resources might be required especially during the data entry phase. A thorough capacity needs assessment should precede the development and establishment of an AIMS. To ensure that the benefits of introducing an AIMS are sustainable, careful consideration needs to be given to minimising staff volatility by ensuring adequate salaries can be paid and costs related to training and skills' retooling met.

Third, data collection must be jointly conducted by both donors and partner governments. For AIMS data to be reliable, they should be provided in a coordinated manner by donors and line ministries. This requires frequent communication between governments, donors and implementers. Government agencies and donors should routinely validate project data and financial information entered. The process of verification is critical to ensure that all data entered are reliable and capture the entirety of activities and funding sources available. Successful implementation of AIMS hinges on making the process of providing data as simple and time efficient as possible. Technology helps in providing easy reporting mechanisms to facilitate data provision. Where donors have strong project information management systems of their own, it may be possible to automate part of the process of data provision and updating. Donor practices should be harmonised and a common framework of disbursement procedures and reporting requirements should be established. Congruence of data between national and local realities should be established. Hence, there should be clarity at all levels on what type of reports the data are going to generate, for what purposes and for which audience.

Fourth, AIMS are costly. The major capital costs are the application purchased or developed, which can range approximately from a few thousand (locally developed databases or spreadsheets) to three hundred thousand US dollars. Governments that have integrated AIMS within the Ministry of Finance have usually benefited from the already existing information systems environment supporting the Public Financial Management Systems and have found that the costs related to the implementation of AIMS can be more easily absorbed within the Ministry of Finance than in other government agencies. Additional capital costs are the equipment and the networks to support the AIMS application. Operational costs include updates, maintenance and human resources and training.

Fifth, there are technical issues such as potential users of the system need to be involved in the design and customisation from the beginning. The tool needs to be designed around existing business processes which can trigger a review of existing business processes and in such a case implementation should be interrupted until objectives and business processes are clarified. It is recommended to keep the number of data entry fields limited. In this context, one should also consider how many different people would be required to provide the requested information. The data quality might be fairly low if a data entry module requires too many different people within a development partner agency to provide information as it is unlikely that the partner agency will enter the data online. Automatic exchange of aid information collected by different donor systems should be pursued to the fullest extent possible. Furthermore, development of a comprehensive web-based AIMS requires a considerable amount of time; and the development time can be cut short through adapting an existing system. Internal or locally available technical capacity, for application code development, user support and help desk functions, networking and internet access and troubleshooting, must be available to maintain and support the chosen AIMS in the long term. This will ensure that the system is maintained up to date and its users are properly trained.

A review (UNDP, 2010) of AIMS for Sierra Leone, Burundi and Central African Republic finds that the there is a need for promoting greater transparency and accountability, and AIMS are constrained by absence of centralized systems of public financial management. There is a felt need for building capacity in recipient governments for establishing and managing AIMS.

Concluding Remarks

The effective use of AIMS requires that: systems respond to existing needs (promoting efficiencies rather than generating additional demands); routine needs of 'end users' are served as opposed to simply producing more data; use of local capacity is maximised; and the system and its developers allow real stewardship of the tool (*i.e.* ownership without responsibility is meaningless) (Gabriel *et al.*, 2006).

AIMS take time to operationalize and therefore the AIMS implementation should focus on the data management required for medium to long term development purposes (UNDP 2010). Information on humanitarian aid might be excluded or maintained as a separate data set. Humanitarian evaluations require sensitive information, for instance, protection information, and there is a heightened need for confidentiality. Constraints on data gathering may be increased during evaluation (Bonino 2014). AIMS may include data on humanitarian aid, but should not be structured according to the same logic or be aligned with development partners internal systems, as this would make it even harder to align aid information management to government systems when they emerge (UNDP 2010).

To conclude, AIMS are an essential but not sufficient tool for monitoring and evaluating humanitarian resource allocation. It is important that AIMS are aligned with human rights considerations. Furthermore, both humanitarian and non-humanitarian activities need to be viewed together. Information on both humanitarian and development assistance will facilitate development planning and monitoring particularly at the local level; and allow assessment of sustainability of humanitarian interventions.

References

Accascina Gabriel, Aidan Cox, Jörg Nadoll, Dasa Silovic, Brian Hammond, and Rudolphe Petras (2006). Role of Aid Information Management Systems in Implementing the Paris Declaration on Aid Effectiveness at the Country Level, DAC Working Party on Aid Effectiveness, OECD

Bonino, F. (2014). Evaluating protection in humanitarian action: Issues and challenges, ALNAP Working Paper, London: ALNAP/ODI

Hallam, H. (1998). Evaluating humanitarian assistance programmes in complex emergencies, Relief and Rehabilitation Network Good Practice Review No. 7, London: ODI

Michael Bamberger and Marco Segone, How to design and manage Equity-focussed evaluations? http://mymande.org/sites/default/files/EWP5_Equity_focused_evaluations.pdf, UNICEF, accessed in December 2014

Richard Heeks (2006). Implementing and Managing eGovernment – An International Text, Sage Publications Ltd.

Solthis (2014). Managing Risk in Fragile States: Putting Health First! http://reliefweb.int/report/world/managing-risk-fragile-states-putting-health-first-optimising-efficiency-global-fund-s, accessed in December 2014.

UNDP (2010). Comparative Experience: Aid Information Management Systems in Post-Conflict and Fragile Situations, October, United Nations Development Programme, New York, USA.

Chapter 16

People: The Real Source of Information and Insight: Evaluation Methodology and Challenges in Fragile Conditions

Based on the Experience Gained at Earthquake Reconstruction and Rehabilitation Authority (ERRA) as Chief Knowledge Management during 2006 and 2008

Khadija Javed Khan*

President,
Pakistan Evaluation Network (PEN), Islamabad, Pakistan
E-mail: kkhan01.kjk@gmail.com

ABSTRACT

In the aftermath of 2005 massive earthquake, a huge programme with a financial outlay of approximately US$ 5 billion with the support of international community and national, bilateral and multilateral partners was launched covering the nine districts with 3.5 million affected population.

The Earthquake Reconstruction and Rehabilitation Authority (ERRA) gained insights and learned quite a many important lessons by conducting its first programme review in 2006. It was not a stand-alone exercise, but part of a comprehensive system devised by the Knowledge Management team to reach out affected communities and bring back information, insights and relevant suggestions from the people to improve upon the delivery of the programme.

Despite the physical challenges of travelling to distant areas, the programme review team benefited from the interaction with stakeholders, partners and communities in the field and built a wholesome picture of the situation including 9 district baselines, two programme reviews and 4 sectoral case studies. Although it was difficult to meet the

---
* *Current Residence*. Schuettaustrasse 48/7, 1220 Vienna, Austria

expectation of the people in fragile conditions, however, the frequent contact created a mutually trusting working relationship with them.

The Programme Review 2005-2006 was an experiment to drastically change the outfit of an official review process and documentation by introducing unedited articles and feedback from the community, the executive summary of the World Bank Mission as well as financial information and performance gaps in the delivery of the programme.

Introduction

One of the key concerns in evaluation is to identify authentic sources of information which not only enrich the data but also provide insights into the issue to make it contextually and programmatically relevant; therefore enhancing the probability of its being utilized and useful in contributing towards improvement. One may ask what could be the most valuable source of information that an evaluator looks for; and the answer is **'people'**. Particularly in humanitarian crisis situation, the affected communities, though hard to reach, are the most reliable source of information as they are faced with real life situations every day, night, hour, minute and second of their life in uncertain, fragile and constantly changing conditions.

In October 2005, a large part of northern Pakistan and Pakistani side of Kashmir was jolted by a huge earthquake of an intensity that measured 7.6 on the Richter scale, resulting in tremendous loss of life and property, destruction of natural resources and with it means of peoples' livelihoods. An estimation of damages is given in the Table 1 for quick reference.

Within a short period after the first emergency and early recovery interventions, the government established the Earthquake Reconstruction and Rehabilitation Authority (ERRA) to start work in the affected areas. In 2006, ERRA developed 9 district baselines and conducted its first programme review, and subsequently in 2007, the second programme review as well as a number of sector specific case studies to assess the progress of work, its efficiency, effectiveness and sustainability and most of all, to strengthen the ownership of communities and stakeholders.

Table 16.1: Estimation of Earthquake Damages

Damages Caused by the Earthquake 8 October 2005	
Number of People Died	73,338
Number of People Injured	128, 304
Families affected	500,000
Population affected	3.5 Million
Area affected	30,000 Sq. km
Educational institutions destroyed	6,298
Health units destroyed	796
Houses destroyed	600,000
Roads damaged	6,440 km
Services - Telecommunication,Power, Water and Sanitation went out of function	50-70 per cent

Source: ERRA/UN Early Recovery Plan, May 2006 (Updated by ERRA in September 2006).

The response continuum consisted of four phases; in which ERRA's role was critical to build upon the first three phases for maintaining momentum and sustainability in the long term.

Table 16.2: Response Continuum

Phase 1 *Immediate*	Phase 2 *Short-Term*	Phase 3 *Mid-Term*	Phase 4 *Long-Term*
Rescue and relief operations;	Sustaining population and displaced persons	Early Recovery Operations	Reconstruction and Rehabilitation
Crisis management;	Supplementing local response capacities		
Damage			
Assessment and control;	Revival - civil administration and		
Maintenance and restoration of Infrastructure.	essential services		

ERRA's programme was by far the most challenging post-crises undertaking due to its geographical and socio-economic scope and financial commitment of around US$ 5 billion.

The Knowledge Management Cell that was tasked to conduct programme reviews consisted of a team of 6 professionals that included 1 Programme Evaluator (Team Leader), 4 Researchers and 1 GIS specialist, supported by two short term external consultants. The team was directly reporting to the CEO to ensure its independence.

ERRA Programme Review

Methodology

A systematic inquiry[1] was carried out, following the evaluation ethical principles of relevance, transparency, independence, stakeholders' participation, cultural sensitivity, rigor, accurate reporting and do no-harm[2]. The team developed multiple methods to collect data and further added an innovative approach to demonstrate its authenticity by inviting direct unedited information in the form of articles based on personal experiences of partner organizations from among the Civil Society Organizations in the field.

Nature of Data

Both **quantitative** and **qualitative** data was obtained from in-house sources and directly from stakeholders/partners encompassing programme related information and cross cutting issues.

---

1   Guiding Principles for Evaluators, American Evaluation Society, 1994 (revised 2004); http://www.eval.org/p/cm/ld/fid=51

2   DFID Ethics Principles for Research and Evaluation, Final Version July 2011, https://www.gov.uk/government/uploads/system/uploads/attachment_data/file/67483/dfid-ethics-prcpls-rsrch-eval.pdf

Approach

Participatory approach was applied during all the phases of review. The main emphasis was on reaching out as many stakeholders from among the community as possible for interviews, focus groups, meetings, workshops and joint reviews of specific sectoral programmes as well as site visits complemented by the documentary and visual evidence of progress, efficiency and effectiveness of the programme. At the same time, finding gaps and documenting issues highlighted by stakeholders or observed directly by researchers.

Tools

☆ **Both desk and field research** was conducted to prepare district baselines.

☆ **In-house collection of data** on sectoral progress and qualitative information was facilitated by the computer section and media section, respectively.

☆ **Field visits** were carried out to **directly collect data/information** from the sources through **semi structured interviews, focus groups, meetings, workshops, site visits** and briefings by managers and stakeholders at district and local level.

☆ Information was backed by **documents, interview notes** and/or **visual evidence** such as photographs and videos.

☆ **ERRA programme managers, executives and relevant district government officers were invited to write their experiences** in specific areas of expertise/sectors.

☆ **Stakeholders/Partners in the field were invited to write their experiences** and contribute data, information, and articles including recommendations for the improvement of the programme.

Planning and Execution

In the outset, ten steps to results based monitoring and evaluation[3] were kept in mind by the research team and pursued during the process, particularly, obtaining agreement of stakeholders to conduct a review with their cooperation.

Scope of Programme Review

Besides the soft issues, the scope of the programme review covered the following eight sectors:

1. Education
2. Health
3. Livelihoods
4. Water and Sanitation
5. Housing, Shelter and Camp Management

---

3 Jody Zall Kusek and Ray C. Rist, Ten Steps to a Results Based Monitoring and Evaluation System, the WB 2004.

6. Needs of Vulnerable Groups
7. Governance and Disaster Risk Reduction
8. Common Services and Coordination

Three cross-cutting themes were mainstreamed in all interventions *i.e.* social and economic rights, gender equality, and environmental sustainability.

Data Collection and Organization

The data was collected from inside and outside sources; the main focus remained on acquiring information directly from people affected by the disaster who could give an accurate account of three key aspects:

☆ What existed before the earthquake in the locality and in what condition?

☆ What has been destroyed and to what extent?

☆ And what needs to be rebuilt to turn the disaster into an opportunity in order to realize ERRA's vision of 'Build Back Better with Hope and Dignity'.

An overview of sources of data/information is given below that resulted in compiling the following main products.

☆ 9 District Baselines/Profiles[4] (8 documents as one document combined two district profiles)

☆ 2 Programme Reviews 2005-2006 and 2006-2007 [5]

☆ 4 Sectoral Case Studies[6]

Analysis and Reporting

This comprehensive and diverse data/information set was compiled progressively and fed into the next exercise. For example, the **District Baselines** were the first set of information that provided basis for the programme review, followed by sectoral case studies to gauge programme effectiveness in selected areas.

ERRA was already maintaining a comprehensive database with all the necessary quantitative information and a hotline for receiving feedback from the public to identify and address problems in real time; therefore the **Programme Review** focused on qualitative aspects of programme delivery vis a vis its sectoral strategies. Instead of using numbers first, the researchers used the qualitative information from the field and supported the same with numbers, graphs and visual evidence as well as perceptions of the people.

The report lay out was designed as follows:

The first part consisted of the programme overview, immediate response, challenges and programme vision followed by sector specific progress vis a vis its

---

4   http://www.erra.pk/publications/publications.asp

5   http://www.erra.pk/Reports/ERRA-Review-200506.pdf

6   http://www.erra.pk/publications/publications.asp

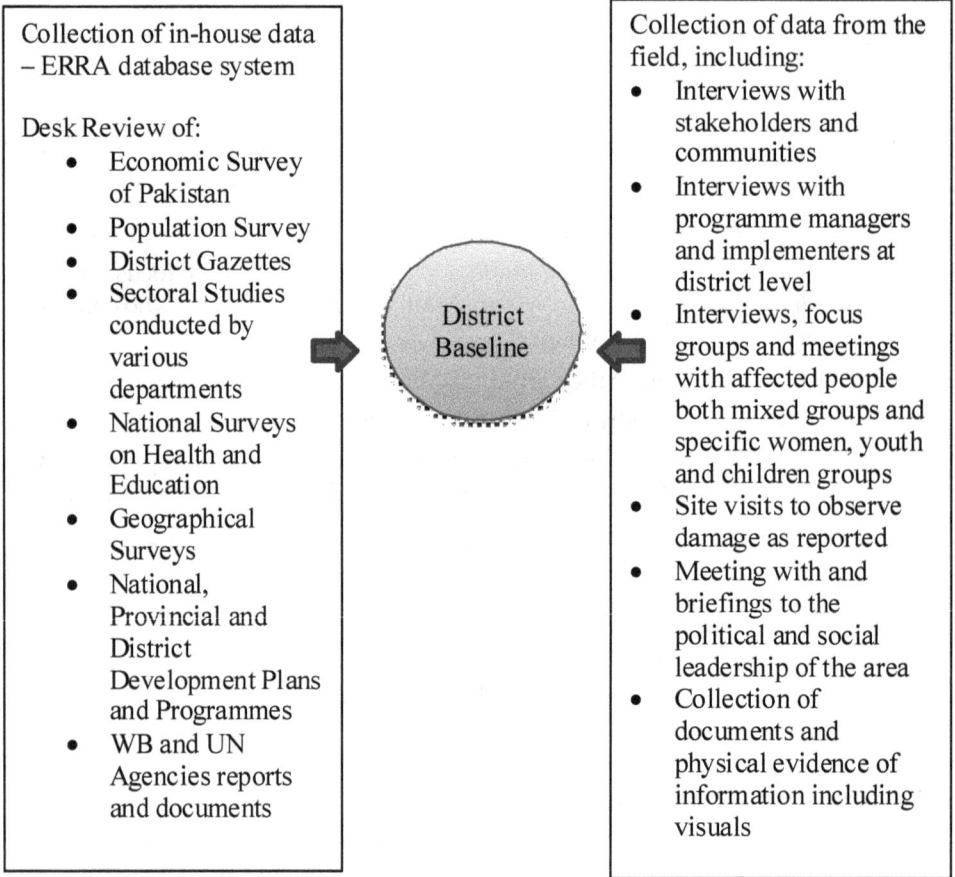

| Collection of in-house data – ERRA database system

Desk Review of:
• Economic Survey of Pakistan
• Population Survey
• District Gazettes
• Sectoral Studies conducted by various departments
• National Surveys on Health and Education
• Geographical Surveys
• National, Provincial and District Development Plans and Programmes
• WB and UN Agencies reports and documents | District Baseline | Collection of data from the field, including:
• Interviews with stakeholders and communities
• Interviews with programme managers and implementers at district level
• Interviews, focus groups and meetings with affected people both mixed groups and specific women, youth and children groups
• Site visits to observe damage as reported
• Meeting with and briefings to the political and social leadership of the area
• Collection of documents and physical evidence of information including visuals |

Figure 16.1

strategy and plan of action including qualitative issues as emerged in the field and noted during the implementation, monitoring and review process. This part also included chapters on core management processes such as ERRA leadership, monitoring and evaluation, sponsorship/partnerships and financial management.

The second part consisted of unedited direct input from stakeholders and partners including UN agencies, and civil society and community organizations who reflected upon their experiences and lessons learned.

The third part (Annexes) consisted of an overview of financial cost of reconstruction and executive summary of the WB Review conducted by a multi-disciplinary mission of experts.

Main Results

Qualitative

The above exercise resulted in drawing attention of all concerned including the government authorities, international and national partners, stakeholders, media

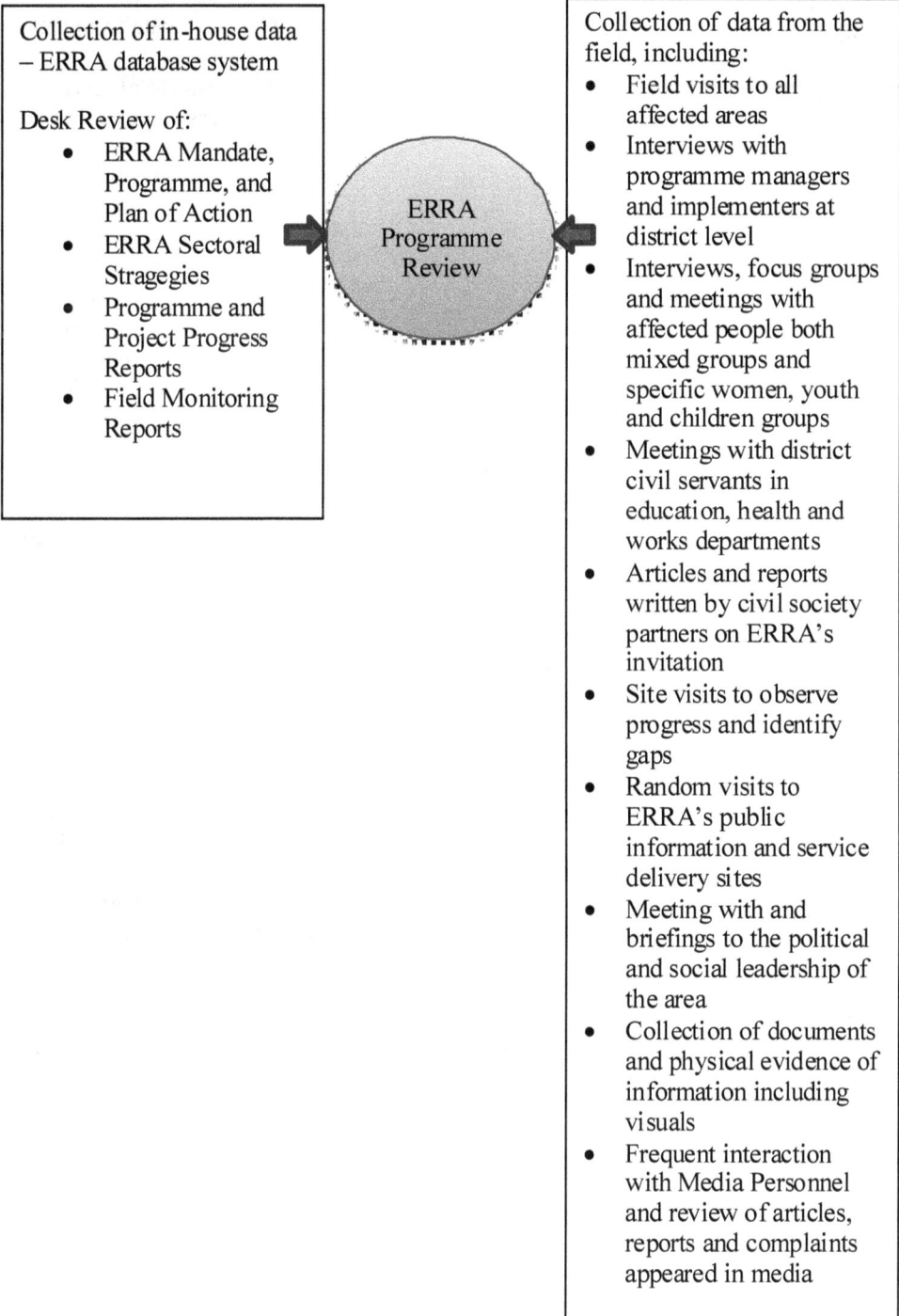

| Collection of in-house data – ERRA database system

Desk Review of:
• ERRA Mandate, Programme, and Plan of Action
• ERRA Sectoral Stragegies
• Programme and Project Progress Reports
• Field Monitoring Reports | **ERRA Programme Review** | Collection of data from the field, including:
• Field visits to all affected areas
• Interviews with programme managers and implementers at district level
• Interviews, focus groups and meetings with affected people both mixed groups and specific women, youth and children groups
• Meetings with district civil servants in education, health and works departments
• Articles and reports written by civil society partners on ERRA's invitation
• Site visits to observe progress and identify gaps
• Random visits to ERRA's public information and service delivery sites
• Meeting with and briefings to the political and social leadership of the area
• Collection of documents and physical evidence of information including visuals
• Frequent interaction with Media Personnel and review of articles, reports and complaints appeared in media |

Figure 16.2

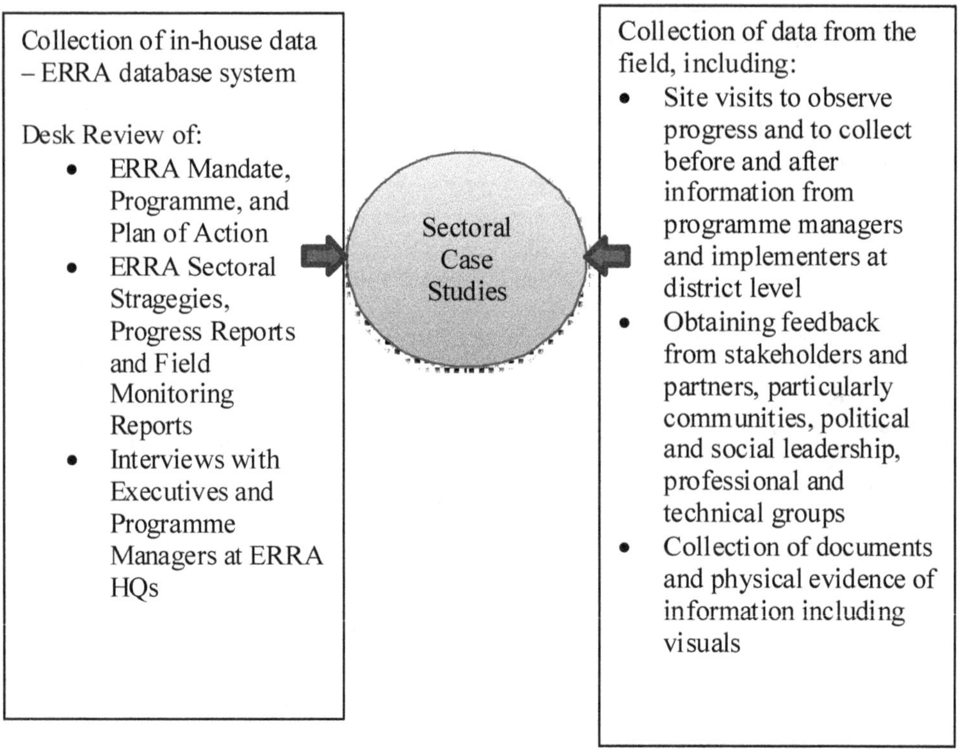

| Collection of in-house data – ERRA database system

Desk Review of:
• ERRA Mandate, Programme, and Plan of Action
• ERRA Sectoral Strageqies, Progress Reports and Field Monitoring Reports
• Interviews with Executives and Programme Managers at ERRA HQs | Sectoral Case Studies | Collection of data from the field, including:
• Site visits to observe progress and to collect before and after information from programme managers and implementers at district level
• Obtaining feedback from stakeholders and partners, particularly communities, political and social leadership, professional and technical groups
• Collection of documents and physical evidence of information including visuals |

Figure 16.3

and the general public to the huge challenge faced by the country and the nation as a whole in the aftermath of the earthquake. While highlighting the short term achievements, it also revealed gaps in the strategy, delays in the implementation of certain segments of the programme and the delivery and quality of services as well AS funding gaps. The criticality of the engagement of stakeholders in the programme and their ownership strongly emerged during the process and was well acknowledged by ERRA.

Quantitative

A glimpse of quantitative performance is given below as an example. Whereas the common services and coordination was effectively delivered, the implementation and delivery rate in health, livelihood and housing remained far short of plans and expectations of the people. Similarly, support to vulnerable groups and service delivery in education, water and sanitation lagged behind the schedule (Figure 16.4).

Challenges

There were two top sensitive challenges:

1. How to manage **expectations of the people** who were devastated having lost their loved ones, homes, possessions and livelihoods.

Funding Overview of the Early Recovery Plan (as of 31 August 2006) *							
Sector	Total Cost (S)	Avaiable Funds ($)	Funds to be Identified ($)	Precentage Funded	Expenditure (S)	Imple-mentation Rate	Delivery Rate
Education	37,774,180	29,766,763	8,007,417	79%	14,341,859	48%	38%
Health	36,967,496	22,161,683	14,805,813	60%	3,954,667	18%	11%
Livelihoods	96,955,073	61,590,917	35,364,156	64%	13,629,595	22%	14%
Water and Sanitation	25,264,600	13,976,610	11,287,990	55%	9,179,472	66%	36%
Housing Shelter Camp Management	32,015,023	16,865,184	15,229,838	53%	2,932,238	17%	9%
Support to Vulnerable Group	10,081,079	4,634,450	5,349,052	46%	2,838,575	61%	28%
Governance	8,328,604	6,237,132	2,091,472	75%	3,469,857	56%	42%
Common Services and Coordination	18,808,564	16,339,560	2,469,004	87%	11,803,560	72%	63%
Total	266,194,619	171,572,299	94,604,742	64%	62,149,823	36%	23%

*Preliminary, based on near-complete data
Source: ERRA/UN

Figure 16.4

2. How to **obtain accurate data and information**[7] to plan, strategize and deliver the programme.

Moreover, the practical challenges faced by the research team were unique to the phases of evaluation as listed below:

Constructing District Baseline

☆ **Access to and availability of data and information**: Due to the fact that most of the government office building and their record was destroyed in the earthquake, proxy estimation using various secondary sources of data was made by a team to prepare baselines of affected districts.

☆ **Validity of data:** The team had to travel to remote areas in the affected districts to contact relevant local people including government officials, political leaders, community organizers, entrepreneurs, workers and households to verify various segments of information *e.g.* number of schools that existed prior to the earthquake and the number needed to be constructed in view of demographic changes during the last many years as well as needs impacted by the earthquake.

☆ **Tabulation and Analysis:** Designing uniform tables and conducting analysis was difficult due to gaps in data and the differing structures of information being obtained from various sources.

---

7   Ref. Presentation in George C Marshall Centre, UN OCHA Desk Officer on UN perspective; http://www.erra.pk/Reports/ERRA-Review-200506.pdf; p.15.

Programme Review

☆ **Scope of Review**: Considering the geographical and social scope, financial outlay, sector diversity and technical delivery mechanism of the programme, the task was too large to be completed by a small team of professionals without enlisting support of stakeholders at the head office, in district offices and at the local community level.

☆ **Interaction in the Field:** The team was constantly in action, either travelling in the field, meeting people, collecting information, synthesizing, writing, adding visuals and putting everything in order for verification and finally clearance by the CEO. During this activity, the most important concern was to stay within the mandate and not to go overboard to commit something to people on behalf of the organization or say something outside the policy line. This was peculiar due to fragile conditions and expectations of the people faced with adversity. Every visitor to community was considered a messenger of hope.

☆ **Writing and Synergizing the Review:** The programme review included contribution from various stakeholders complementing the core set of information acquired from the in-house database. The main challenge was to create synergy in the various parts of the report and to ensure accuracy and consistency.

☆ **Acceptance of the report:** Some of the colleagues in ERRA who came from hard core bureaucratic background were not fully appreciative of including a segment on the perception of communities based on the contribution from civil society non-profit partner organizations in the field. Only the positive signal from the CEO could save the work of the research team.

☆ **Printing of the Review**: Working with the printers was another challenge as the staff at the printing firm was not familiar with technical terminology used in the review, and the research team had to work 24 hours 7 days a week with them to ensure accurate and timely production. The proof reading and editing consumed a lot of time and efforts.

Case Studies

For the purpose of developing and compiling case studies, an international expert was engaged who was provided access to information and to communities for collecting and synthesizing information. There was no major challenge except that the progress was slow and there was not sufficient evidence to demonstrate major achievements. However, the case studies also contributed to the research experience of ERRA.

Lesson Learned

The main lesson learned from this exercise is cited in the title *i.e.* **'People are the most authentic source of information'** in any evaluation process. Reach out to them and benefit from their direct feedback, insight and even criticism.

Chapter 17

Real Time Evaluations: One NGO's Experience of Learning and Improvement of Humanitarian Responses

Vivien Margaret Walden

OXFAM
Oxford, UK
E-mail: vwalden@oxfam.org.uk

ABSTRACT

A rapid onset humanitarian crisis is often difficult to both monitor and evaluate as the situation changes quickly and it is imperative to address the immediate needs of the affected population. At the same time, a process is needed for response teams to be able to review what has been achieved, what the gaps are and what changes are needed in order to make an impact. The Real Time Evaluation (RTE) methodology has been called "a snapshot in time" or a chance for results of the evaluation to immediately generate learning and change. The choice of a good evaluator who acts more as a facilitator with good communication and analytical skills is key, as is the feedback of results in real time and the results consensus between response team and evaluator. The RTE methodology in theory lends itself to better learning as it is facilitative and participatory. However, agencies should not make the mistake of using the RTE methodology as a substitute for the full programme cycle of monitoring and evaluation as aspects such as impact are not part of the RTE process. The use of the word evaluation may be misleading as the true real time process is more about learning and adjustment than it is about judging a programme or measuring the impact. The Real Time Evaluation was never meant to be the magic tool but as one of many evaluative processes that enhance learning in humanitarian situations.

Keywords: *Evaluation humanitarian learning.*

Background

A **humanitarian crisis** is an event or series of events which represents a critical threat to the health, safety, security or wellbeing of a community or other large group of people, usually over a wide area. Armed conflicts, epidemics, famine, natural disasters and other major emergencies may all involve or lead to a humanitarian crisis that extends beyond the mandate or capacity of any single agency.[1]

A rapid onset humanitarian crisis is often difficult to both monitor and evaluate as the situation changes quickly and it is imperative to address the immediate needs of the affected population. Access may be difficult especially in conflict areas. Programme staff will be preoccupied with implementation and may struggle to find time to engage in a review. Baseline data are often missing or are unreliable. Even if baseline data are collected, the fact that the target population is traumatised needs to be considered; space to grieve must be respected. Reliable secondary data may be hard to obtain. When evaluations do take place, they are often ex-post or at such a protracted time after the immediate response that changes to implementation are not feasible. Thus a gap was identified for a more action-oriented evaluative process that would identify changes that need to be made "in real time" as well as contribute to organisational and the wider sector learning.

The Real Time Evaluation (RTE) methodology has been called "a snapshot in time" (Walden *et al.*, 2010) or an exercise that is "carried out whilst a programme is in full implementation and almost simultaneously feeds back its findings to the programme for immediate use" (Sandison 2003). The concept became popular in the 1990s when several UN agencies adopted the methodology: UNHCR in 1999, after criticism of their response in Kosovo, introduced "rapid analytical evaluations" in the early stages of a response. Also in the same year, the organization for Economic Cooperation and Development/Development Cooperation Directorate (OECD/DAC) recommended carrying out an evaluation during the early response to an emergency in order to assess and to make adjustments accordingly (OECD/DAC 1999). As one NGO who adopted the methodology, Oxfam carried out its first RTE in Darfur and Chad in 2004 but without using pre-agreed benchmarks.

A Real time Evaluation is not "simply a cut-down version of a conventional humanitarian evaluation" (UNHCR 2002). The characteristics of a RTE can be summarised as: timeliness, interactivity and perspective (UNHCR 2002). The process is timely in that it is undertaken in the early phase of a response – Oxfam, for example, carries out their RTEs within the first six to ten weeks after the onset of an emergency. Interactivity means that evaluators are involved in the planning process (feeding into programme change) and perspectivity is achieved by evaluators being able to look at different levels (country, regional office and headquarters) as well as bringing experience and learning from other responses. Evaluators can be experienced external consultants, from headquarters or for example for some NGOs from the organisation's field offices but external to the programme being evaluated. The findings and recommendations should be made available before the evaluation team leaves the

---

1   http://humanitariancoalition.ca/info-portal/factsheets/what-is-a-humanitarian-crisis

country during an interactive debriefing or what Oxfam has termed a day of reflection. Another hallmark of a RTE is that there is a set of pre-determined criteria or benchmarks against which progress is measured. An example of these is the seven benchmarks used by Oxfam adapted from various sources including UNHCR:

1. The response meets the agreed standards for speed and timeliness and is judged in comparison to other actors

2. Relief provided is appropriate to the context, based on expressed needs of both women and men and is of a quality and scale that meets the organisation's commitments and expectations

3. An effective management structure is in place, providing clarity and well-communicated decision-making and direction

4. Key support functions are sufficiently resourced and effectively managed. Security is well managed and risks that are being taken are calculated and documented

5. Internal relationships are well coordinated and mutually accountable under the leadership of the country team and the organisation is having a positive influence on other actors through mechanisms such as the cluster system

6. The program has considered the longer term implications and has taken connectedness into consideration

7. Campaigning, advocacy, media and popular communications, or a combination of these tools, are appropriate for the context, are well executed and are having a positive influence on the response

The Advantages of Using the RTE Methodology

The choice of evaluators or facilitators is crucial: having an understanding of the organisation's mandate and ways of working means that time is not lost in having to familiarise oneself with the agency. Having a facilitator who understands the system and who has themselves been part of a previous response may help to build trust and foster a climate of mutual respect – although the opposite can also be true if the evaluator is not a respected member of the organisation.

For Oxfam, the evaluators are typically drawn from other country teams or from head offices, and therefore are conversant with Oxfam policies and procedures reducing the briefing time to specific programme context. The team leader should ideally have previous experience of conducting an RTE and thus can bring the lessons learnt from a previous response to the current one. Timing of an RTE is crucial: not too early when teams are still trying to respond to immediate needs but early enough to be able to make changes to the activities and approach if needed. The methodology encourages participation of the affected population and the opinions of women and marginalised groups such as ethnic minorities are especially sought even if this means making an extra effort to set up meetings in a venue and at a time that is convenient for them. For conflict areas where access is impossible, the use of technology such as Skype or mobile phones has shown that communities can still be involved, although triangulation is necessary and care taken to ensure that the people being interviewed are really who they say they are. The experience of Oxfam in both Afghanistan and Somalia has shown that community views can be sought, albeit in limited numbers.

The findings and conclusions are presented at a day or reflection held in country where staff and local partners are invited to discuss findings and to agree on reccomendations: the so-called results consensus approach (Guba and Lincoln 1989). An action plan is also developed with responsiblities for actions defined. The adventage of this day is that the report simply becomes a record of proceedings and a protracted period of agreeing on findings and recommendations is usually avoided. Learning should be acted upon in "real time."

The Disadvantages of Using the RTE Methodology

One of the risks with an RTE is that the limited time period may mean that wider consultation with the affected population may not be possible and therefore the views of the communities may not be fully represented. It takes a skilled evaluator to ensure that this either does not happen or that the risk is minimised. The methodology relies heavily on good interviewing techniques, the ability to triangulate the information and good analytical skills: something that not all internal evaluators have. Unlike more conventional summative evaluations, the RTE evaluator should be more facilitative in approach, working along side organisational staff and partners to assist them in making sense of the situation and identifying mutually accepted improvements. Some RTEs can throw up unpleasant issues that need to be handled carefully if the team is to accept failure and rectify their mistakes.

Since 2002, RTEs have grown in popularity and have been carried out by most large NGOs, the UN agencies and even some donors. Not all of these were in truth "real time" as especially the larger UN processes were taking place several months after the onset of the crisis. Any evaluative process held in the first six months was being labeled "real time" and other ex-post or summantive evaluations were not being commissioned as a result. RTEs do not look at impact or lasting changes that have occured in the lives of the affected population: this is not their purpose.

Does an RTE Lead to Better Learning?

Oxfam has carried out three reviews of all RTE reports since taking up the evaluative process for all large emergency responses. Two of these looked at the process itself as well as the results. In 2008 an online survey was conducted among 50 members of staff who had participated in an RTE. There was a 48 per cent response rate and out of those 54 per cent had found the process useful. The results had been used by 56 per cent to make immediate changes to the programme and 74 per cent said they would use the results to make changes for the next phase of the response programme.

The 2010 review was a qualitative desk study but again, "the general conclusion reached from the survey and desk review was that RTEs were considered useful and did lead to improvement in programming in the majority of cases" (Oxfam 2010). It was felt that the selection of the right people for the evaluation team was crucial and that having team members who were "open and honest makes all the different between a good and bad RTE." It was also felt that that by focusing on "improving the quality of the response by encouraging people to think about change; clear and concise recommendations coming out of RTEs make it possible for teams to follow up and bring about these changes."

The 2013 review of RTEs showed a mixture of improvements and recurring areas for improvement. It would be easy to draw the conclusion that programme staff do not learn from past experiences and that the same mistakes keep getting made but that is too simplistic an explanation. Learning is complex and there is a vast literature on how people learn. In humanitarian programmes there is high staff turnover where knowledge and experience is lost and the culture of not questioning senior managers persists today even in the most enlightened aid agency. RTEs are an opportunity for staff to pause and reflect and as long as they trust the evaluators, they can use the process to express their concerns in way that will not jeopardise their relationship with managers or other staff members. Feedback from evaluation team leaders has been that sometimes interview sessions can feel like a cathartic outpouring of frustration from staff members who feel that their views are not being heard.

In 2014, a comparison was done across benchmarks of the results for programming from a met-analysis of six RTEs conducted from December 2013 to June 2014, with a similar meta-analysis in April 2013, which considered 13 RTEs. There *had* been learning from previous responses: a good example was that complaints and feedback mechanisms were now systematically in place across almost all programmes although there was not always evidence that the feedback had led to programmatic change. An improvement had also been made around clarity of roles and responsibilities and the impact this had on the timeliness of a response. Making comparisons across qualitative reports is always difficult but it was possible to track some positive trends not just in the same country but across countries and regions.

Conclusions

The RTE methodology in theory lends itself to better learning as it is facilitative and participatory. If the process comes at an early stage in the emergency response and is led by an evaluator who understands the facilitative approach rather than judgement by an external team, there is a good chance that teams will learn from the process and make changes to areas that have shown to be weak. The advantage of using internal staff is that they can take the experiences from one programme to another (such as lessons from the tsunami being used in a similar disaster in the Philippines). The disadvantage is that not all staff are able to adapt their knowledge to a new context and the recommendations made may reflect more of their home-country situation than that of the programme under evaluation.

The process is especially useful in conflict and fragile states where the situation can change rapidly. A more conventional evaluation may be commissioned but by the time the evidence has been analysed and presented, the situation has changed and the information becomes obsolete. The Real Time Evaluation methodology can be adapted to be a light touch but rapid way of assessing what has been done and what needs to improve.

In recent years, the question has been raised around the usefulness of the RTE and the over-reliance on this type of evaluation. Interestingly, the United Nations opted to rename their evaluations as reviews rather than Real Time Evaluations (Telford 2009), as did the Disasters Emergency Committee (DEC) in the UK. The use of the word evaluation may be misleading as the true real time process is more about learning and adjustment than it is about judging a programme or measuring the impact. Experience in Oxfam has shown that the development of an evaluation policy and minimum standards for monitoring, evaluation, accountability and learning are encouraging a more realistic programme cycle that includes RTEs, learning events, after-action reviews and summative evaluations. The Real Time Evaluation was never meant to be the sole magic tool but one of many evaluative processes that enhance learning in humanitarian situations. Maybe the time has come to change the name across the sector.

Acknowledgements

The opinions expressed in this paper are those of the author and do not necessarily reflect those of Oxfam. The paper is based on meta-analysis reviews and surveys conducted by Oxfam during the past ten years. The author would like to thank Peta Sandison for reviewing the paper.

References

Guba, EG and Lincoln, YS. 1989. *Fourth Generation Evaluation.* Newbury Park: Sage Publications

OECD/DAC. 1999. Guidance for Evaluating Humanitarian Assistance in Complex Emergencies. Available at: http://www.oecd.org/derec/dacnetwork/35340909.pdf

Oxfam. 2010. Real Time Evaluation Review Report: A meta-evaluation of 12 Oxfam International Real Time Evaluations, Oxford: Oxfam

Sandison, P. 2003. Desk review of Real-time Evaluation Experience, available at: http://www.unicef.org/evaldatabase/index_29680.html

Telford.J. 2009. Review of Joint Evaluations and the Future of Inter Agency Evaluations. Available at: https://docs.unocha.org/sites/dms/Documents/Review_of_joint_evaluations_and_future_of_IA_evaluation[1].pdf

UNHCR. 2002. Real Time Evaluations. Available at: www.fmreview.org/FMRpdfs/FMR14/fmr14.17.pdf

Walden, V.M. Scott, I and Lakeman, J. 2010. "Snapshots in time: using realtime evaluations in humanitarian emergencies", *Disaster Prevention and Management: An International Journal,* 19 (3): 283 - 290.

Chapter 18

Research Methods for Understanding and Measuring Impact among 'Hard-To-Reach' Undocumented International Migrants in India

Nabesh Bohidar[1] and Navneet Kaur[2]

CARE India
New Delhi
E-mail: [1]nbohidar@careindia.org, [2]nkaur@careindia.org

ABSTRACT

A large number of migrants from Nepal and Bangladesh migrate to India for work and employment. There is very little information on these populations, as researchers have desisted from focussing one undocumented migrants due to the feasibility of carrying out research among those who may not be recorded in the destination. The EMPHASIS project, implemented by CARE India provided an opportunity to carry out research among these populations.

The paper draws upon the experience of carrying out a baseline and end-line research among hard-to-reach migrants in India. The baseline research explored various facets of the migrant's lives including the vulnerabilities that they faced, apart from benchmarking indicators for the evaluation. The end-line research focussed on examining outcomes and impact of the intervention through a quasi-experimental design.

The examination of the methods used in the research indicate that feasibility of carrying out a quantitative survey has major challenges and wherever possible it is better to utilise project lists as sampling frames. Secondly, researchers need to be mindful of sensitive questions, related to migration, current status as well as violence, and use mixed methods. Finally, it would be useful to carry out research at both source and destination sites to construct a more complete picture.

Introduction

Cross border mobility is central to the lives of many Nepalese and Bangladeshis as they move between their countries and India in search of better work and livelihood opportunities. A treaty signed between India and Nepal in 1950, enables citizens of both countries to travel and work freely across the border (Wagle *et al.,* 2011). Without asimilar treaty or policy,manyBangladeshis wishing to migrate, do so without papers, and slip through the porous borders, even though large portions of the border are fenced on both sides and tightly regulated (Sultana *et al.,* 2011). Both Nepalese and Bangladeshi migration to India is undocumented, though the process of migration, the status and identity of migrants in India, differs vastly. Due to the undocumented nature of the migration as well as a relative lack of interest about their vulnerabilities, very little research is available on these populations, more so in India.

There are numerous challenges in carrying out studies on undocumented migrants. It is difficult to identify people who may not be part of official records and would be very thinly distributed across locations. Also due to transient nature of the population, they (need to) move from one location to another in search of employment and an enabling environment. CARE India implemented the EMPHASIS[1] project, in order to understand the characteristics of migration, and to reduce vulnerabilities of migrants and their families to HIV. This paper is based on the evaluation studies carried out for the projectfor measuring impact of its intervention in India[2]. The objective of this paper is to present the methodologies used for carrying out studies among the Nepalese and Bangladeshi migrants in India, and in doing so highlight some of the key challenges in carrying out research among undocumented international migrants.

A robustresearch design that was used in order to first capture the change from the pre-intervention scenario among Nepali migrants (NP) and Bangladeshi migrants (BD) in India (Figure 18.1).

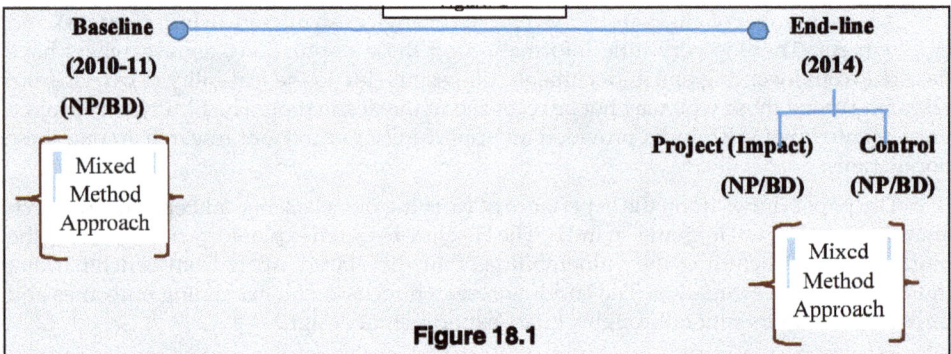

Figure 18.1

---

1   'Enhancing Mobile Populations' Access to HIV and AIDS Services, Information and Support', was a five year project, supported by the Big Lottery Fund, UK. It was a comprehensive project intervening at source, transit and destination sites across two international routes between Nepal-India and Bangladesh-India.

2   Though evaluation studies were carried out in all three countries, only the India section is discussed in this paper.

The Baseline

The baseline research was carried out with the objective of (a) understanding the patterns and drivers of mobility for the purpose of work; (b) to measure and document the main vulnerabilities that contribute to HIV risk; and (c) to understand access to service along the mobility route.The study entailed judicious mix of both quantitative and qualitative methodologies.

Quantitative Study Design for Baseline

The study population consisted of migrants, from Nepal (NP) and Bangladesh (BD) at selected intervention locations in Kolkata, Delhi and Mumbai[3], who were potential impact population[4]using a structured questionnaire. A mapping exercise was carried out prior to the study by CARE India, in which a total of 198 locationsfor NPs and 65 locations for BDs were identified as the universe.

The estimation of sample size was based on the following statistical formula;

$$n = \frac{p(1-p) \times Z^2}{e^2} \times \text{Deff}$$

Taking 50 per cent P value, with e set at 5 per cent, z at 1.96 and Design Effect at 1.2, the sample size was estimated as 456. Further considering 20 per cent as causality, the calculated sample size was 547, which was rounded up to 550.Considering a set of sample for each of the two streams (NP and BD), a total sample of 1100 was taken.

A two stage sampling procedure involved selection of survey locations (SLs) at the first stage, followed by selection of respondents. All mapped locations were grouped in three sub-groups based on high, medium and low concentration of migrant population in selected areas. The required number of SLs,in accordance with the sample size, were selected within each sub-group using the systematic random sampling procedure.At the second stage, a household listing was carried out in selected SLs and respondents were selected.

Qualitative Methodology for Baseline

In order to understand the context further, and to complement the quantitative survey, a qualitative study on vulnerability of migrantswas carried out. Qualitative methodology entailed administration of in-depth interviews (44) and focused group discussions(14) across migrants who lived with and without their spouse.

Strategies, Challenges and Learning from Baseline: What Worked Well and what Did Not

Identifying the impact population posed a big challenge during baseline. The migrants (especially BDs) do not want themselves to be identified as such, due to the

---

3   Kolkata included selected locations in North 24 Parganas; Delhi included selected locations in National Capital Region; and Mumbai included selected locations in Thane.

4   Migrants (in the age group of 15-49 years) who have come from Nepal and Bangladesh to India for the purpose of employment.

resultant harassment, and legal action that may ensue as a result. Therefore the EMPHASIS project devised certain 'markers' for Bangladeshi migrants in India, which were utilized till the end of the project. These markers werebased upon dialect, religion, peer and key informants. The BD impact population was referred to as "Bengali Speaking Population" and not Bangladeshis, by the project. (EMPHASIS, 2014)

At the design stage the impact population was narrowly defined. The initial cut off for the impact population was those that have stayed in India for at least 3 months and to a maximum of 5 years (Figure 18.2). However, getting correct responses to the amount of time spent in India/location became difficult. Themigrants perceived

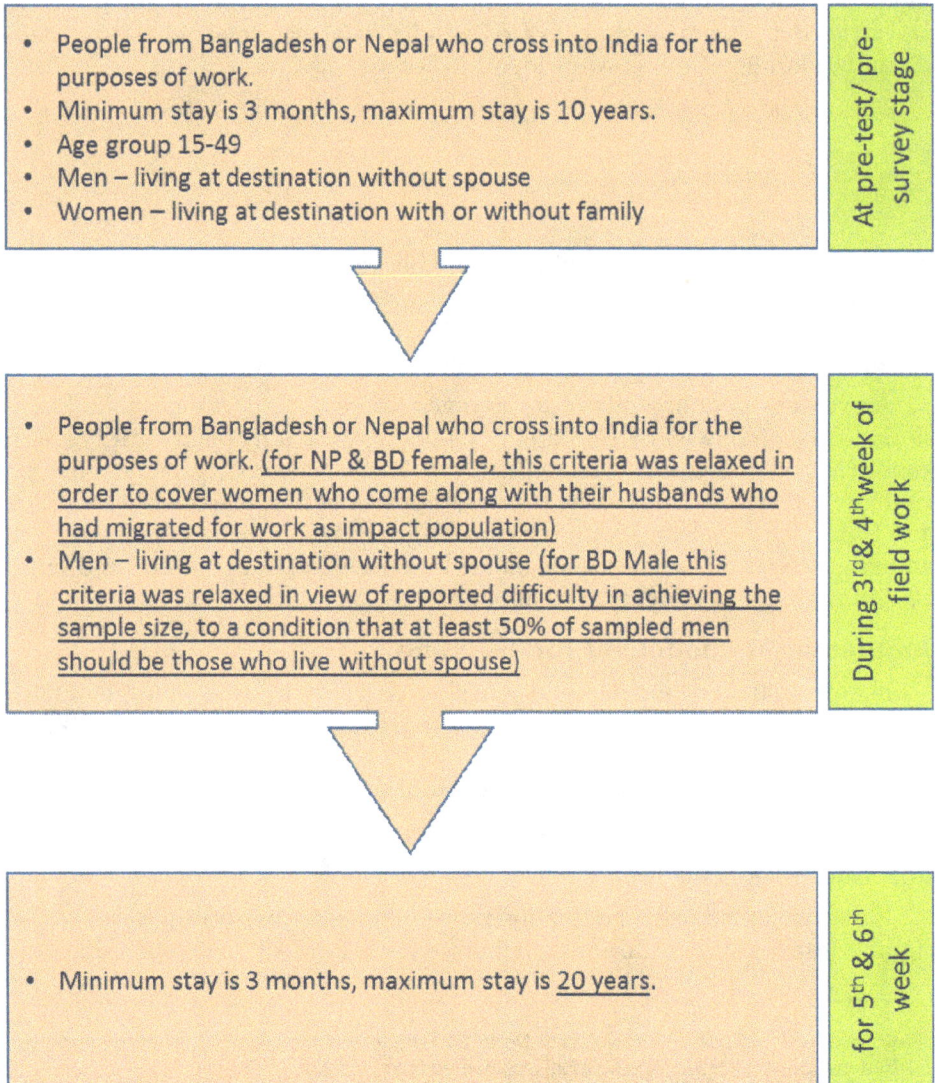

- People from Bangladesh or Nepal who cross into India for the purposes of work.
- Minimum stay is 3 months, maximum stay is 10 years.
- Age group 15-49
- Men – living at destination without spouse
- Women – living at destination with or without family

At pre-test/ pre-survey stage

- People from Bangladesh or Nepal who cross into India for the purposes of work. (for NP & BD female, this criteria was relaxed in order to cover women who come along with their husbands who had migrated for work as impact population)
- Men – living at destination without spouse (for BD Male this criteria was relaxed in view of reported difficulty in achieving the sample size, to a condition that at least 50% of sampled men should be those who live without spouse)

During 3rd & 4th week of field work

- Minimum stay is 3 months, maximum stay is 20 years.

for 5th & 6th week

Figure 18.2: Definition of Impact Population.

benefits of being considered as residing in the same place for a longer time, increased their chances of getting access to services and identity (through ration cards and voter identity cards). For instance in a sub-locality in Mumbai (which according to the key informants had come up in the last 3 to 4 years) had respondents saying that they have been staying in the same place even before that. Thus a response bias in terms of duration of stay resulted in making respondents ineligible and the survey long drawn.Finally, the cut off period for a migrant in India was expanded from a maximum of 10 years to 20 years.

The second criteria (migrated for work) to be relaxedto include similar proportions of females in the study, who overwhelmingly followed their husbands to India. As a consequence, the initial SLs selected for sampling was inadequate and additional SLs were chosen. This in turn meant that instead of respondents being chosen through household listing, most were based on snowballing. Thus total number of people who could be surveyed as per the planned sampling method (from house listing) was only 11.5 per cent among the NPs and 18.5 per cent among the BDs.

Adopting a mixed methods approach, besides leveraging the commonalities between quantitative and qualitative, added flexibility to the analysis rendering it more complete,and provided insight on Nepalese (Wagle, 2011) and Bangladeshi (Sultana, 2011) migrants. These findingswere taken into account by the project implementers in designing specific strategies (Samuels, 2011; EMPHASIS, 2014) for reaching project goals. The Baseline research also brought forth some contradictory findings (Figure 18.3).

The qualitative findings were more in line with the expectations of research teams, so did they find it because they were looking for these? Similarly, was the quantitative finding a truer picture because of 'representativeness' in the sample? Alternatively, was the survey through a structured questionnaire intimidating to an elusive un-documented migrant?

Sl.No.	Finding from Baseline Quantitative	Finding from Baseline Qualitative
1	Migrants at destination don't face problems in terms of access to services, harassment, feel are treated relatively fairly/no different to other workers/migrants.	BDs (because of illegal status) in India feel excluded, face language barriers, are scared to disclose their identity, maintain distance from their neighboursNPs in India were discriminated by landlords, employers, 'are labelled a thief/criminal without justification
2	Migrants didn't face problems during transit, and pass through relatively quickly.	Crossing borders, and bargaining/negotiating with brokers/middle men was an issue for BD and found to be 'humiliating experience'Stories from returnees in BD are that girls/women are harassed, tortured and raped by border people on both sidesNPs are harassed at the border, are asked to see their IDs, are cheated by transporters, are beaten, etc.

Figure 18.3

The End-line

The key objectives of end-line were (a) to measure the overall achievements towards an effective and integrated cross border model of HIV prevention, care, treatment and support of individuals at the impact population level; and (b) to compare key changes in time to knowledge, attitudes and practices questions between the baseline, and end-line control locations.

Quantitative Study Design for End-line

The quantitative study anchored by a respondent survey followed a robust and rigorous quasi-experimental approach, allowing key indicators to be measured for comparisons between end-line and baseline impact populations[5] and purposefully selected end-line control populations. Control locations were identified within the intervention districts that were similar to EMPHASIS locations and have presence of an estimated migrant population. The control locations were distant (2 to 6 km) from EMPHASIS locations to avoid possible spillover effects of the intervention. Due to the transient nature of the study populations, it was not possible to revisit the same respondents as those interviewed during the baseline activity. Hence the study utilized propensity score matching analysis, which provided an additional level of robustness to the assessment of impact

A two sample formula was used to estimate the sample size, at each stratum level, to allow for meaningful comparison longitudinally between the baseline and end-line; and at the cross-sectional level between an end-line control and impact population:

$$n = D[(z_\downarrow \alpha + z_\downarrow \beta)^2 * ((P_\downarrow 1 (1 - P_\downarrow 1) +$$
$$P_\downarrow 2 (1 - P_\downarrow(2)))/[(P_\downarrow 2 - P_\downarrow 1)]^2]$$

The estimated level of the key indicator to be measured as a proportion at the end-line is 0.5 (P1 =0.5), which is the most conservative assumption. To be able to detect a change, increase or decrease, of 20 per cent P2 is set at 0.6. A confidence level of 95 per cent (Zα = 1.645) and a power level of 80 per cent (Zβ = 0.840) was used. Design effect was taken as 1.2, which was utilized during the baseline evaluation[6]. Using the sample size formula above, these parameter values result in a sample size of 363 respondents per strata. Allowing for a 20 per cent non-response rate results in a sample of 436; rounded up to 440 per stratum. Using the eight stratum listed above, a required sample size of 3,520 respondents is considered.

The experience of the baseline study was taken into account for anticipating the non-response rate at 20 per cent. This was increased to a maximum of 40 per cent, after a pilot respondent identification exercise. However, the actual survey carried out using a focused strategy for respondent identification, including repeat visits,

---

5   Impact population means – BD and NP migrants who were registered by the project.

6   The utilization of a design effect of 1.2 is a limitation of the sample design due to resource constraints.

resulted in non-response rates, which were lower at 25 per cent for BDs among the impact population and (3 per cent) for the control population[7].

For the impact population, the first stage sampling of clusters was based on available impact population lists[8]. Using probability proportional to size sampling (PPS), clusters were sampled in each of the 4 impact population stratum. From the selected clusters, a simple random sampling (SRS) of the required number of respondents were drawn. To further minimize the non-response rate, any individuals listed in year 2012 or before were removed from the sample frame. This sampling method resulted in a self-weighted sample at the each stratum level (Magnani, 1997).

Control locations with estimated population of potential respondents *i.e.* migrants, that has similar socio-economic characteristics were taken as the universe. Asample of clusters was selected using PPSfor each of the 4 control population stratum.Further, to identify specific respondents, a rigorous random walk strategy using multiple segmentation techniques was utilized with appropriate screening questions to identify target respondents.

The findings from the end-line research overwhelmingly supported the achievement of project goals and concluded "The EMPHASIS project, when measured by its key project and outcome indicators, which align with the program log-frame goals, is overwhelmingly successful. These findings are robust when comparing baseline to end-line impact groups (longitudinally) and end-line control to end-line impact groups (cross-sectional) with traditional t-tests, propensity score matched average treatment effects and propensity score matched average treatment effects of the treated. The project has achieved an effective and integrated cross border model of HIV prevention, care, treatment and support, and the data repeatedly demonstrate this accomplishment. Theimpact population respondents consistently cite; lower risk behaviours; increased knowledge of HIV and AIDS; better health seeking behaviours; and EMPHASIS or its associated workers, village health workers, or peer educators are regularly cited as the source of this information. The impact population community environment which has been strengthened and enabled by EMPHASIS for cross-border mobile populations, when evaluated on the bases of people's knowledge, attitude and practices, is strong and growing. This is particularly the case for females and Bengali speaking populations; a population that has a dearth in rights and entitlements within India at the moment" (Ravesloot and Banwart, 2014).

Key Learnings

Conducting research among undocumented migrants is difficult from a methodological perspective. The first and foremost is the feasibility of carrying out a quantitative survey, when it can be difficult to locate them (McAuliffe, 2013) or as in EMPHASIS where migrants (BDs) want to stay hidden. In such cases, it may be

---

7   Control respondents were identified using a random walk strategy which resulted in significantly lower non-response rate

8   These were the "First Meet" lists from routine monitoring data, which listed all impact population in a geographical location.

important to build sampling frames over a period of time through project workers and utilize them[9]. Thus, while respondents refused to reveal their origins to the Baseline survey team, they were more comfortable in doing so to the end-line team. This method will suffer from intervention bias, it can enable a quantitative survey. When sampling frames can be obtained from official records, they have been utilized with varying degrees of success (Smith, *et al.*, 2011).

As the topic of research is sensitive for migrants, who at the destination, are exposed to various pressures, it is more likely that the answers would be in line with what is socially acceptable (Nederhof, 1985). In the baseline research, we have come across differences of responses between the qualitative and the quantitative in relation to living conditions and discrimination. Further, response biases are seen especially when migrants in need of official ids can provide responses that they perceive will strengthen their case for official papers or identity cards.The findings from the baseline indicated that more respondents at the source country talked about facing problems (discrimination and violence) during transit than among migrants who were interviewed in India. Thus there is need to create certain level of comfort to talk about and also exploring those sensitive issues through qualitative data collection methods seems more suitable.

The transient nature of migrants at destination is a challenge for conducting baseline and end-line studies for measuring impact. Routine data collected through the EMPHASIS project indicated a 30 per cent to 40 per cent change in migrant population in one location over the period of a year, though the overall numbers may not change very much. In this context the methodology adopted during the end-line, which considered current migrants from among the "first meet" lists resulted in successful location of respondents. Further,identifying control locations, through criteria that are similar to intervention sites can be successful.

Conclusion

Studies among migrants in the destination is critical as they are current migrants and provide valuable understanding of their vulnerabilities and aspirations. It is possible to utilize sampling frames from interventions, with adequate "do no harm standards". Researchers should be mindful of sensitive questions, which may produce socially acceptable responses, motivated responses and those that may require rapport and trust. Further, a "project-control" design may be preferred over a "before-after" design, in case of transient populations and mixed methods can be used to assess different facets of complex outcomes or impacts, which will yield a broader, richer portrait. Finally, as evidenced through a comprehensive intervention approach, which reached migrants and their families across a mobility continuum (source, transit and destination), with information and services as required by the impact population and not only on health and HIV, a comprehensive research design, which studies migrants and their families at both source and destination provides a more comprehensive picture.

---

9   Adequate "do no harm standards" needs to be rigorously followed.

References

EMPHASIS 2014. EMPHASIS Learning Series. CARE EMPHASIS Regional Secretariat: Kathmandu.

Magnani R (1997). Sampling Guide: Food and Nutrition Technical Assistance Project (FANTA). Academy for Educational Development. Washington.

McAuliffe, M. 2013. Seeking the Views of Irregular Migrants: Survey Background, Rationale and Methodology. Irregular Migration Research Program: Occasional Paper Series. Department of Immigration and Border Protection.

Nederhof, A. J. 1985. Methods of Coping with Social Desirability Bias: A Review. *Eur. J. Soc. Psychol.*, 15: 263–280.

Ravesloot, B. and Banwart, L. O. 2014. CARE EMPHASIS End-line Survey Report.

Samuels, F., M.N. Zarazua, S. Wagle, and M.M.Sultana, 2011. Vulnerabilities of Movement: Cross-Border Mobility between India, Nepal and Bangladesh. Policy Brief. Overseas Development Institute, London.

Smith, P., A. Cleary,M. Jones,S. Johnston, P. Bremner, J. Brown, and R. Wiggins, 2011. A Feasibility Study for a Survey of Migrants. Report Prepared by Ipsos MORI and the Institute of Education for the UK Border Agency.

Sultana, T, A Das, M.M. Sultana, F Samuels and M.N. Zarazua (2011). EMPHASIS Baseline – Vulnerability to HIV/AIDS: Social Research on Crossbordermobile Populations from Bangladesh to India, EMPHASIS: CARE Nepal.

Wagle, S, N. Bohidar, F. Samuels, M.N. Zarazua and S. Chakraborty (2011). EMPHASIS Baseline – Vulnerability to HIV and AIDS: Social Research onCross Border Mobile Populations from Nepal to India, EMPHASIS Regional Secretariat, CARE Nepal.

Chapter 19

Relative Impact of Disaster on Women: An Assessment of 2013 Disaster in Uttarakhand (India)

Senior Adviser and Regional Director,
Ramana Group, Dehradun, Uttarakhand
E-mail: rsgoyal52@gmail.com

Background

It is recognized worldwide that people's vulnerability to risks to a large extent depends on the assets they have. Owing to manifestation of existing gender inequalities in most societies, women often tend to have limited access to assets, in comparison to men. These also limit their capacity to adapt to difficult circumstances. Consequently, women tend to experience far adverse consequences in the wake of natural disasters.

Gender perspective has however not been adequately addressed in disaster research, planning and management. Despite significant progress in integrating gender issues analytically and in the field, neither governmental agencies nor NGOs have as yet fully integrated gender relations as a factor in disaster vulnerability and response, nor have they engaged women as equal partners in disaster mitigation and community-based planning. As Enarson(2000) has argued, *"If addressed at all, gender has been integrated into disaster research and practice as a demographic variable or personality trait and not as the basis for a complex and dynamic set of social relations."*

Seeing disasters "through women's eyes" raises issues for planners, identifies critical system gaps, and brings gender centrally into development and all disaster management related works. Some of these are as below.

☆ What social indicators best predict the relative impact of natural disasters on women and men?

☆ In diverse environmental, political-economic, social, and cultural contexts, how do gender relations differently shape the impacts of natural disasters and (often) varying responses of women and men to these? What cross-hazard and cross-cultural patterns can be identified?

☆ How and to what extent socio-economic development affect women's and men's vulnerability to hazards and their relative ability to recover? How can these patterns be assessed in specific contexts?

☆ To what degree higher vulnerability of women to hazards been included in the design and implementation of emergency response, relief, and reconstruction policies? What has been the effect of the same?

☆ What are the specific short-term needs (in the context of their particular vulnerability) of girls and women in specific contexts? What are women's long-term interests in reconstruction?

☆ What are women's short- and long-term needs as primary household preparers, long-term care-givers, employees and volunteers?

☆ What organizational or other barriers limit response to these needs, under what circumstances, and with what effects?

☆ Are women (which women?) at the table in the development of disaster-resistant communities?

In June, 2013, the state of Uttarakhand (India) has witnessed one of worst disaster in Indian history causing widespread damage to human life and property due to flash flood. About 5000 people have lost lives. Loss of physical property and animal husbandry was several folds higher. Vast tracts of agricultural lands have either been washed off or covered with thick pile of debris and silt, their homes have either been damaged or destroyed, their cattle have been lost, their savings have been carried away in the deluge and many are left with very little or no money. Further, majority of the families in the disaster affected region were depended on tourism, pilgrimage and subsistence farming for their livelihood. As there is little hope of revival of both pilgrimage and tourism in near future and, income from agriculture is too meager to support the family, hardship will haunt the masses for a long time to come.

Women were larger sufferer in these flash floods. Besides the economic and physical losses, many have lost their husbands and children in the prime of their youth. This duel tragedy has shattered the lives of these women and as per newspaper's accounts even lost the desire to live.

This disaster event presents an ideal setting to carry out an in-depth analysis of the effect of disaster on women. The findings could also supplement the rehabilitation efforts being put in by Government and other agencies.

Objectives

The specific objectives of this assessment were as follows.

☆ In the environmental, political-economic, social, and cultural contexts of Uttarakhand, what was the relative impact of June 2013 disaster on women and men?

☆ How did gender relations (differently) shape the impact of disaster and the (varying) responses of women and men to it?

☆ To what extent and how the development patterns in the region has affected the women's and men's exposure to disaster's impact and losses and, their relative ability to recover and cope with the disasters?

☆ To what extent the particular vulnerability of women to disasters was recognized and included in the emergency response, relief, and reconstruction programmes? And with what effects?

☆ What are women's short and long-term needs as the primary home makers, long-term care-givers, livelihood earner etc., in disasters? Were these addressed in emergency response and rehabilitation programmes after June 2013 disaster?

☆ Whether the women (which women?) were at the table in the development of disaster-resistant communities?

Methodology

The objectives demand that this assessment be based on qualitative data collected from the primary stakeholders *i.e.,* women. It also limits options for adopting a rigorous evaluation design for this assessment. Hence, A descriptive design was adopted.

The primary data were collected from select villages in Rudraprayag district (one of the worst affected district in the wake 2013 disaster in Uttarakhand) nearly 9 months after the disaster. Only those villages which have suffered losses (human life, mules, houses and agricultural fields) in the wake of June 2013 disaster were considered for study. Lists of such villages were obtained from the state's Disaster Mitigation and Management Center (DMMC), Govt. of Uttarakhand and Rudraprayag district's disaster cell. Districts officials also provided detailed information on human and other losses in every villages and (till date) compensation paid by the Government to the victims and their families.

On the basis of this information, villages in study universe were divided in three broad categories.

A. Villages which have primarily suffered human loss
B. Villages which have suffered loss of houses and landed property.
C. Villages which have suffered all kind of losses.

From every category, four villages were selected for study purposes (List of study villages is shown in Table 19.1).

Qualitative tools used includes; participatory appraisal, in-depth interviews and case studies. In every village, one participatory appraisal was carried out. Though these were largely attended by the women members of community, men were also engaged in this dialogue. On an average between 12 to 15 persons participated in a meeting. In addition, a total of 42 in-depth interviews were conducted with the women. These women represented average cross-section of rural community who have directly or indirectly faced the disaster in some form or other. Further, 4 case studies of women who have directly faced the brunt of disaster on themselves/their families were conducted.

Some Specific Feature of Study Area

☆ Uttarakhand is a relatively less developed state of the Indian Union. Poverty level in the state is 39.6 percent (proportion of people living below poverty line) which is significantly higher than the national average of 27.5 percent (PHD Research Bureau; 2013).

☆ Despite poor economic status, the state ranks higher than many other states in the country in terms of status of women (assessed on the basis of indicators that include employment status, education, age at marriage, fertility and autonomy).

☆ The socio-economic life in hill districts of Uttarakhand (about 90 percent rural) is largely depended on agriculture, remittances (from the people employed in army, police or other occupations away from the home) and tourism.

☆ In the hilly terrain, though most households own land, land holdings are very small and plots are scattered far and wide. Further, agriculture is largely depended on rains for irrigation. Consequently, the agriculture is of subsistence type and the yields are barely sufficient for the consumption of household alone.

☆ Secondary sector of economy is just nonexistent in the hill districts. As an off shoot, blue collar employment opportunities are very limited and a good number of people choose to migrate to other places for earning livelihood and better economic prospects. Defense sector has traditionally been one of the largest employers for the people from the region.

Remittance/money order economy supports a large number of families in this area.

☆ The Rudraprayag, Chamoli and Uttarkashi districts house most scared hill pilgrim shrine of Hindus and Sikhs. Badrinath, Hemkund Sahib, Kedarnath, Gangotri and Yamunotri are visited by over one million people every year. Religious tourism has emerged as a major contributor to the economy of these hill districts as well as the state. It is estimated that the religious tourism annually contributes over Rs. 150 billion to state exchequer and at the same time sustains livelihood of more than 2,00,000 families.

☆ Mountain villages in the region have a higher proportion of women (to men) as compared to plain areas. It is largely attributed to out-migration of people for employment.

☆ By and large women in the hills are hard working and toil for long hours. Apart from routine household chores, they take care of the agricultural pursuits as well. A women in a rural household typically works for 12 hours a day, of which 3.5 hours are spent in gathering fodder and fuel from the forest, while 3.5 hours are spent on livelihood related work that includes agricultural works (that are largely managed by women) and, 4.75 hours are spent on household related work (Chopra and Gosh; 2000).

☆ With respect to status of women, it was noted that women play an important role in decision making related to family or household. However, when it came to decision making or expressing opinion on matters related to village or wider community, they had very little say. Women have relative freedom of movement but with the consent of husband or elders in the family. Among harizans, however, women were not expected to go out alone.

☆ In June 2013 disaster, major loss of human lives and property was incurred in the settlements in Rudraprayag district, particularly in villages located around Kedarnath shrine and or the banks of Mandakini river flowing down from Kedarnath.

Findings

Human and Physical Losses

A brief comparative profile of the study villages is presented in the Table 19.1. This table also reflects the losses suffered by the villages in the wake of June 2013 disaster.

A close look at Table 19.1 reflects that the study villages were of all dimensions and population sizes. Their total population ranged between 90 to 1700 persons. Further, more than two thirds of villages were depended on Kedarnath shrine for their livelihood. In 5 (38.5 percent) villages, for more than 90 percent families the main source of income was pilgrimage to Kedarnath temple.

Among the 13 villages studied, 9 (69.2 percent) have incurred human losses during the June 2013 disaster. It varied from 54 persons (in Dewali Bhanigram) to 3 (in Railgaon and Damer). All the dead were males, including boys. It is to be noted that in Kedarnath area, women do not accompany their husbands/family members to Kedarnath when they work there. Interestingly, in Badrinath area, it is other way round (there women company their men folk to work).

Table 19.1: A Comparative Profile of Study Villages in Relation to Losses Suffered during Disaster

Sl.No.	Name of Village	Total Population	No. of HH	Per cent HH Depended on Kedarnath	Average Landholding (Nali)	Human Casualties	Mill itch Animals Lost	No. of Houses Damaged	Agriculture Land Lost (Nali)
1.	Dhani	247	46	78	3	5 Males	–	–	40
2.	Badasu	600	156	96	4	23 males	40 ponies	–	35
3.	Kalimath	645	250	80	5	–	11 ponies and 125 sheep	35	30
4.	Dewali Banigram	1254	300	90	7	54 males	7 ponies	–	–
5.	Nakot	503	148	12	4	–	2 ponies	–	–
6.	Silli	300	110	None-directly	2	–	–	40	48
7.	Railgaon	90	15	85	5	3 boys	2 ponies	–	30
8.	Gavni	517	110	None-directly	2	–	–	34	60
9.	Damar	511	207	18	2	3 males	1 pony	–	–
10.	Dunger	500	111	90	1	13 males	5 ponies	–	30
11.	JalTalla	346	58	70	10	20 males	35 ponies and 35 sheep	–	130
12.	JalMalla	625	98	90	7	17 males (3 boys)	40 ponies and 300 sheep	–	60
13.	Trijugi Narayan	1700	250	90	5	20 males	8 ponies	65 shops	–

This practice was discussed during the participatory appraisal meetings and, several reasons were put forward to explain this practice. It was observed that conditions in Kedarnath are tough for long stay, there is no culture of taking families along and, accommodation is also limited. Where as in Badrinath, there is a proper village and most people providing services and carrying out business at Badrinath, have build their houses in the village.

All the boys who died in the calamity were in 12 to 15 years age group. These boys had accompanied their fathers or family members to Kedarnath for assisting them in their work. Their schools were closed for summer vacations and they had no other work at home. It has been noted that it was a general practice for the boys in this area. Many boys also worked as casual workers at Kedarnath to earn money during pilgrimage season.

The loss of houses and agricultural land was largely in the villages which were located on the river bank or hit by the land slide. The lost ponies were in service at the time of disaster and were washed off in the flood.

Emergency Relief and Compensation for the Losses

The June 2013 disaster was one of the major tragedies causing a huge loss of life and property. It also inflicted a large blow to the economy of the state. As happens after disaster, national and state governments, NGOs and other organizations rushed in with supplies to provide emergency relief to the affected people.

If the stories narrated by the people are true, so many supplies were provided and stockpiled by the people that it could easily last for 6 to 12 months for every family in the affected areas. Even during the course of the fieldwork of this study (April 2014; 10 months after the disaster) supplies were pouring in. Though there were usual complains of misappropriation, loss in transit, not reaching to right people and the like, it must be accepted that every affected family in the region has received more than adequate emergency ration and other supplies. Also, the emergency support system became operational, within two-three days of mishap.

The second important component of relief services is the compensation for the loss of human lives, animals, home and agricultural land. *Here it must be born in mind that the word compensation should not be taken by its dictionary meaning. No amount of money can compensate for a human life. The financial assistance is just a token of support and given to help the family members (of deceased) to overcome the tragedy and difficult time being faced by them. Similarly, it is very difficult for any Government or other support agency to fully reimburse the cost of losses incurred by people in natural calamities. The financial assistance is generally (should be also be treated as) a token of support to overcome the difficult time and meet the immediate needs. The prime motive behind giving financial support (compensation) is to demonstrate solidarity with the affected people in difficult time.*

Importantly, in a very short time, all the lives lost were accounted for, verified and a significant compensation (Rs. 5 lakh) was paid by the government to the next of kin. In most cases the beneficiary was wife of deceased person. Besides, a reasonable compensation was paid for the animal lost, houses damaged and loss of agricultural land.

Grant for reconstruction of all the damaged houses in one of the villages (Gabni) is being provided under a World Bank aided project. Several other organizations and individuals have also provided cash and other support to the families who have lost a member. In Dewali Bhanigram where 54 people have died, support poured in like anything, though it is an affluent village. *Sulabh International*, a Delhi based NGO has adopted all the orphaned families in the village and is providing them a monthly stipend.

Further, all the damaged schools were repaired, either by government or NGOs and became functional within 2 months after the disaster. Government extended the summer vacations by another one month so that by the time students go back to studies, schools become functional. Some schools are also being rebuilt by the NGOs.

Roads are the life line of people in the mountains. There was a huge damage to roads due to flood and landslides. Importantly, manageable road/trolley/ropeway connectivity was re-established in all the areas in a very short time. However, it would take a while to bring roads to their earlier shape as there are challenges of weather, topography, budget, ownership and many other things.

In subsequent analysis it would be brought forth that all these emergency support, supplies, compensations, reconstruction etc. have had a significant bearing on the women ability to withstand the impact of disaster.

Impact on Women

Women shouldered major responsibility of looking after the family immediately after the disaster. Along with their regular work, they were also pressed in to collecting emergency supplies provided by the government and other agencies. At time they had to walk for long distances to collect these goods. Women headed households or the women who have lost their husbands have faced more hardship in this respect.

Almost all the women have reported that they have received adequate supply of ration from the government or other sources to overcome difficult times. Ninety percent families have received more than two weeks' ration. In case of 40 percent families, supplies lasted for more than a month. 81 percent women observed that these supplies were adequate for their needs. Only complaint was that they had to walk long distances as distribution was made at the fixed locations. About 19 percent women have complained that the distribution of supplies was not equal.

It was also noted that no special provision was made for women in distribution of relief material. They received same treatment or facilities as were given to men. Further, it has often been reported that under distress conditions, women are more vulnerable to exploitation; particularly the sexual exploitation. No such incidence was however reported by any of the respondents.

Loss of source of livelihood is termed as the major adverse impact of disaster by most women (79 percent).

Because of loss of income they are finding it difficult to run the household (78 percent).

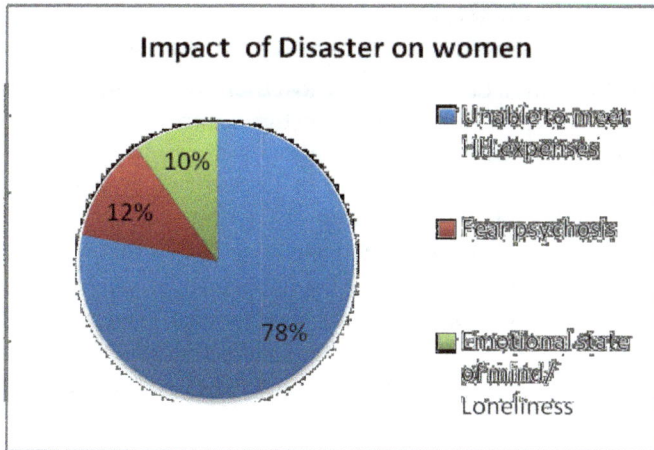

Figure 19.2: Impact of Disaster on Women.

About 12 percent women are still scared of rains and have a constant fear that things may go wrong again. Loneliness and deterioration in emotional security is also reported by about 10 percent of women; particularly by those who have either lost a family member or whose men folk have moved out for alternate employment.

Discussion

Analysis of the information collected for this study creates a scenario (of women) that is quite different from what is normally observed in most post-disaster settings. Though, it is true that in this disaster, both men and women have suffered, but they have suffered differently. The specific vulnerability of women in post-disaster situation as reported in the literature, was, by and large not observed in this study. Gender specific impact of disaster was also molded differently. Let us review the findings to find out possible explanation for the phenomenon observed in the study.

· In the hills of Uttarakhand, there are two parallel economies. One is controlled by women and the other by men. Women till the land, grow food and make a significant contribution towards running the household. They are self-employed and not wholly depended on their husbands for economic upkeep. Men either run small business in the same or neighboring places or work outside and send remittances.

In the study area, economic losses incurred to women (occupation wise) in June 2013 disaster were very small except in villages where their land was washed off. Most women are continuing with their usual occupations.

On the other hand, men have suffered heavily. Most men were engaged in business of providing services to Kedarnath pilgrims. With the disaster, pilgrimage has literally come to a halt and the source of their income has dried out totally. These men do not have any other vocational training. The development pattern in the hills has also not created many other employment avenues for them. This disaster has not only exposed their economic vulnerability but has brought forth a dilemma too. Thanks to women's agricultural activities, the families may not starve, but for long term sustenance of the family, men will have to find alternate employment soon. The chances of rival of

Kedarnath pilgrimage are definitely there, but it may take quite a while. There is a desire to explore alternate avenues, but nobody seems to know what and how of it.

☆ In the study area only men have lost their lives in Kedarnath disaster. The answer could probably be found in the culture or lifestyle of women. Either due to their pre-occupation with agriculture, or due to culture or due to lack proper family accommodation in Kedarnath, women do not normally accompany their men to Kedarnath. Consequently, only men (from the villages covered by present study) have died in Kedarnath disaster.

Loss of men folk has shattered the life of a large number of women in their young age. As widow re-marriage is not very common, it further adds to their misery. Yet there are several positive phenomenons which have saved them from devastation. Firstly, their engagement in agricultural activities keeps them occupied and gives economic stability. Secondly, recognizing their vulnerability, compensation paid by the government for the loss of their men folks was swift and reasonable and most importantly the money was particularly paid in wife's account only. It allowed them to maintain a second source of income and also live a life with adequate means. Thirdly, they were not alone in their sorrow. From every village, several women have widowed at the same time and in the same tragedy. Also, whole society and nation stood by them. Adoption of all the widowed women in Dewali Bhanigram by Sulabh International is a live example of such a support. This support (providing economic stability) also serves an important long term need of women affected by the disaster.

☆ Women's social and physical vulnerability at the time disaster was neither reported nor observed in any of the villages taken up for the study. Perhaps it has much to do with hill culture, where women are given their due status and in all probability they have earned it by putting in their tireless work.

No incidence of sexual exploitation (that could be linked to disaster) was reported in any of villages or in nearby areas. Women were not even prepared to discuss about it.

However, traces of emotional distress were still noted (about 10 percent) amongst widowed women.

Case: Mrs. Vijay Laxmi Nautiyal, Gram Pardhan, Gavni

Mrs. Vijay Laxmi Nautiyal is Pardhan of Gavni Panchayat. She is an active person and has played an important role in rehabilitation of people after the June 2013 disaster. A large number of people (# 34) have lost their houses and belongings in the flash flood in her area. There were no human losses. She had organized their temporary stay in the homes of other villagers. She also worked with administration and donor NGOs to ensure that all the affected people received adequate support and timely compensation for the losses. She even motivated a World Bank team to provide house reconstruction grant (Rs. 5.00 lac) to the affected families in her villages. These houses are being reconstructed (some of these houses are better than earlier ones!).

☆ Supply of ration and items of daily needs is an important short term need of women and the families at the time of disaster. In the study area, timely and ample supply of relief material in all villages and to all the families has played an important role in bringing life to normal soon after the disaster. It was largely helpful and appreciated by women. Perhaps, until unless warranted, special provision for women in emergency relief supplies may not be necessary.

☆ It has been observed that normally women are not involved when it comes to disaster management and mitigation planning and implementation. But there are examples which show that given the opportunity, women are capable of effectively responding to emergencies (case study of Mrs. Manorama Nautiyal). Further, women were also at the forefront in collecting relief supplies from collection centers.

References

Chopra, R and Gosh, D; (2000). "Work participation of Rural Women in Central Himalayas". Economic and Political Weekly. Mumbai, Dec.30.

Enarson, Elaine; (2000). "Gender and Natural Disasters". Recovery and Reconstruction Department, ILO, Geneva. September.

PHD Research Bureau; (2013). "Life Ahead in Uttarakhand: Rebuilding Infrastructure and Reviving Economy". PHD Chambers of Commerce, New Delhi. August.

INNOVATION IN EVALUATION

Chapter 20

Learning and Experience of Using Social Network Analysis in the Context of Indian Civil Society Organization Networks

Samik Ghosh[1], Aniruddha Brahmachari[2]
and Atanu Ghosh[3]

[1]*Programme Coordinator;* [2]*Manager,*
– Monitoring Evaluation Learning, Oxfam India
[3]*Assistant Professor of Economics*
Bankura Christian College, University of Burdwan,
West Bengal, India
E-mail: [1]*samik@oxfamindia.org and samikghosh_crj@hotmail.com;*
[2]*aniruddha@oxfamindia.org;* [3]*iipsatanu@gmail.com*

ABSTRACT

In recent years, development sector organizations have moved a step further in connecting other organizations working on similar issues or who has the potentiality of working on similar issues. Social Network Analysis of development partners in this context plays an important role in visualizing the pattern of network and the network strength. But the term "Network" too often is used only as a metaphor; that is, only a figure of speech to present an image. Social network analysis has emerged from an abstract mathematical theory called "Graph Theory" but in this paper we have tried to explain the matter in a simplified way to suggest and share learning with contemporary social development professionals on how social network analysis can be carried out by using suitable statistical package and also how to interpret the findings in the social development context. Social network analysis is a large and growing body of research on the measurement and analysis of relational structure. Here, we have discussed the fundamental concepts of network analysis, as well as easier methods currently used in the field of development research.

Issues pertaining to data collection, analysis and interpretation of findings are discussed. This paper also speaks about fundamental concepts and terminologies related to social network analysis.

Keywords: *Network, Relationship, Social network analysis, Social structure.*

Introduction

The term *social network* was used for the first time in 1950 in *sociometrics,* the science that seeks to obtain data on social behavior and to analyze it. The latter introduction of mathematical tools and computing enhanced the evolution of Social Network Analysis (SNA) and Analytics. The mathematical basis of SNA arose out of the fields of graph theory, statistical and probability theory, game theory as well as algebraic models. In fact, it was from these theories, especially graphs, that the Internet and various virtual networking concepts were derived. It is quickly widening multidisciplinary areas involving social, mathematical, statistical, and computer sciences (Burt, Minor and Associates,1983; Wassermann and Faust,1994; Dutta and Jackson, 2003).

In today's development research and process oriented programming a frequently used term is 'Network'. Civil Society Organizations (CSOs) frequently use references to personal networks, community networks, organizational networks, and institutional networks. However, the term is used only as a metaphor; that is, only a figure of speech to present an image. Rarely, CSOs reflect that networks are actual relationships and linkages that can be measured with both quantitative and qualitative matrix.

A social network is made up of individual organizations (technically known as "nodes") that are tied (technically known as "tie") with one or more other organizations through certain kind of relationships. The term *social network* refers to the articulation of a social relationship, ascribed or achieved, among individuals, families, households, villages, communities, regions, and so on.

Social network analysis provides a research perspective and methodology by which the structure of a particular environment can be assessed, thereby assisting in determining the state of cooperation amongst organizations. It provides visual maps of the linkages between organization and institutions. These types of analysis allows us to understand networks overall and their participants. Not only does network analysis provide visual maps, but it also allows for mathematical analysis of these maps. For example, at the network level, we can assess the degree of interaction between network members by calculation of the network density or calculate the degree to which a network is highly centralized (all or most connections are to one or few members).

The overall structure of this paper is as following. After a brief introduction of social network and Social Network Analysis (SNA) we have discussed about how SNA is different from conventional statistical analysis used in development research. Following this we have given example of graphical representation of network and discussed the meaning of networking including basic terminologies related to SNA.

We then discussed about the importance to study of a network and how SNA helps social development planner to formulate a better plan through involving more development partners. Finally, we concluded that without having rigorous knowledge on mathematics and statistics how social development professionals can perform SNA and use its result for better social inequalities.

Social Network Analysis: Different Paradigm other than Conventional Statistical Analysis – Why and How?

Most social scientists apply wealthy knowledge of basic univariate and bivariate descriptive and inferential statistics. Many of these tools find immediate application in working with social network data. There are, however, two quite important distinctive features of applying these tools to network data.

The most important feature, social network analysis is about relations among actors, not about relations between variables. Most social scientists have utilized their statistics knowledge with applications to the study of the distribution of the scores of actors (cases) on variables, and the relations between these distributions. We learn about the mean of a set of scores on the variable "income." We learn about the Pearson zero-order product moment correlation coefficient for indexing linear degree of association between the distribution of actor's incomes and actor's educational attainment.

The application of statistics to social networks is also about describing distributions and relations among distributions. But, rather than describing distributions of attributes of actors (or "variables"), we are concerned with describing the distributions of relations among actors. In applying statistics to network data, we are concerned the issues like the average strength of the relations between actors; we are concerned with questions like "Is the strength of ties between actors in a network correlated with the centrality of the actors in the network?" Therefore in network analysis, we focus on relations, not attributes.

Graphical Representation of Network

To understand the term "Social Network" it is important to understand the term "Network". The example below would be beneficial to understand the meaning of network. Network analysis for Civil Society Organizations (CSO) means how individual CSOs or a group of CSOs are interrelated to each other. For example in the diagram (Figure 20.1) out of 8 CSOs, CSO1 to CSO5 are interconnected with each other. Suppose CSO-1, CSO-2 and CSO-3 are sharing information among themselves and they have created a small subgroup of information sharing network (subgroup-1). On the other hand CSO-4 and CSO-5 are also sharing information between them and created another network as subgroup-2. Similarly CSO-6 and CSO-7 have created another network. But one of the members CSO in subgroup-1(CSO-2) and in subgroup-2 (CSO-4) is sharing information between themselves. Therefore we can say that subgroup-1 and subgroup-2 are connected to each other through two bridging organizations (CSO-2 and CSO-4). But the members of the third subgroup are not connected to any other network members; therefore they are called as "Isolated network". In the diagram CSO-8 is not keeping relationship with any other CSO or

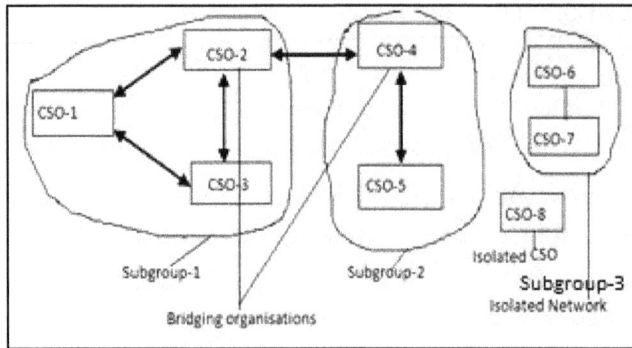

Figure 20.1

network and it is called as Isolated CSO. Thus network analysis gives us a pictorial representation regarding how the CSOs are interrelated to each other and what are the important bridging organizations connecting different networks.

Like CSO network presented in the diagram there could be different types of other network, such as electrical networks, computer networks, neural networks, telecommunication networks, transportation networks and of course, social networks. Our focus of this paper is on social networks. A social network is any type of relational ties (relationship) or links between individuals, groups, organizations, and institutions. Relational ties or links can be such things as friendship, exchange of information or money, or seeking advice. In our example we have talked about relationships on information sharing among CSOs. Thus, an actual social network is the structure or map of how different people, groups, organizations or institutions are connected together based on a certain type or types of relationships. Social network analysis in the context of CSOs, help us to identify the structure of a "network" of actors (individual CSOs), the quality of relationships in it, identify gaps in relationships, and find those organizations that are best positioned within the network to cooperate and serve as an advocate on behalf of the network. In our example above CSO-2 is in a better position to advocate on behalf of subgroup-1 and subgroup-2.

Why it is Important to Study a Network?

A very well known thought in the social sector monitoring and evaluation is that "you can't change it, if you can't measure it". Therefore to improve the network strength in terms of its function and credibility it is important to understand the pattern of networking amongst CSOs. In a sense, we want to understand the network so that we can "weave" (can create) it into a better network. Understanding the network requires knowledge (refer to Figure 20.1): the connection among the network subgroups (subgroup 1 and 2 in example); identification of isolated networks (CSO-6 and 7); isolated members (CSO-8); who are the central members (CSO-2) here because it is in the larger network and playing as a bridging role in connecting other networks); and who are the "bridges" (CSO-2 and CSO-4). In the given example we have discussed about three network sub groups only. But in practice there are huge numbers of networks working on different issues and having different expertise.

Why to Improve or Weave a Better Network?

The basic reason is that when networks of organizations are better connected they are more cohesive, productive and resilient. Also, the denser the network, the easier it is for information to spread and coordination to occur. However, the network should have not only internal connections but also some degree of external connections to get new information and access different types of resources. For example (refer Figure 20.1) network subgroup 1 is a national network which is working on inequality issues and network subgroup 2 is a global network working on the same issue. If both these networks are connected to each other then it would be easier for the network one to raise the national issues at global level with the help of the global network (subgroup 2). Now suppose the isolated network (*i.e.* CSO-6 and 7) is the network of global donor agencies, if this isolated network can be linked with the existing two other networks who are working on the inequality issues at national and global level then such linkage would improve the strength of the existing network through financial support. In short, expanding the network for improving its strength is called weaving network.

So far we have discussed about the evolution of network analysis, meaning of social network, few terminologies related with social network analysis, graphical representation of network, how social network analysis is different from conventional statistical analysis carried out on social development indicators. Now we will briefly discuss how to conduct social network analysis and how to interpret findings using the experience of OXFAM India on conducting network analysis among the CSOs under the Wada Na Todo (WNTA) network (data represented here doesn't reflect views of Oxfam views, authors used this model as only for research interpretation purpose).

The Steps of Network Analysis: Design, Data Analysis and Representation

For conducting network analysis among the CSOs under WNTA network the study involved the following steps:

a) **Study design:** How data was collected (personal interview, online questionnaire, telephonic interviews etc), what were the information collected (the scope of the study, *e.g.* CSOs may be linked to each other through information sharing, resource sharing and advocacy) and from whom information will be gathered.

b) **Study tools and data collection:** In this CSO network analysis example given here, an online version of questionnaire was designed using online Survey Monkey questionnaire tool designing software and the survey monkey questionnaire web link was shared with the WNTA network CSOs and partner CSOs were asked to respond through online submission using specific web link attached with an introductory email.

c) **Creating a database:** In survey monkey link responses were stored in MS-Excel format. After downloading data from survey monkey link restructuring of data from vertical format to horizontal format prepared. Participant organizations were coded as per their characteristics. For example being a national level network WNTA was coded as "NN33", which means there are at least 32 other national networks whose information was obtained in the survey. Once that is done, data are numerically coded and converted into *.txt file (Dershem, L., T. Dagargulia, L. Saganelidze, S. Roels, 2011). From the raw excel data two types of data sets were prepared in *.txt format, Node data with node properties and tie data. Explained below data types,

   i) **Node data:** Name of the CSOs who have participated in the survey and the CSOs who have not participated but they were linked with CSOs who have participated (as responded by participant CSOs)

   ii) **Node properties:** These were the characteristics of the nodes. In our study, this could be local CSO, international CSO, national network, regional network, global network, global donors etc.

   iii) **Tie data:** Tie data represented the linear relationship of one CSO with others. CSOs may be tied up with each other through information sharing, resource sharing or through advocacy. Tie data showed on which particular aspect the organizations were linked to each other.

d) **Drawing the network:** The analysis may be carried out using different softwares. Many of them are freely available on web. The examples are Gephi, Netdraw, UCINET etc. In the current analysis we have used Netdraw software. Node and tie data together in a single *.txt file is imported in Netdraw software (http://www.analytictech.com/netdraw/netdraw.htm) and analysis was carried out.

Analyzing the Network

Looking at the inter-linkages among different network subgroups the strength and capacity of the network is analyzed. The example below shows the information sharing network among CSOs under WNTA network. The network map presented below gives us an idea for information sharing how the organizations are connected to each other. For conducting joint advocacy it is very important to share information amongst the organizations. The network map below shows 26 distinct networks. Out of these 26 networks 16 are inter connected with each other through 7 different bridging organizations. These bridging organizations play an important role in connecting different networks. Information sharing holds centralization in network interaction. For advocacy purpose identification of such bridging organizations are very important. These bridging organizations are as following;

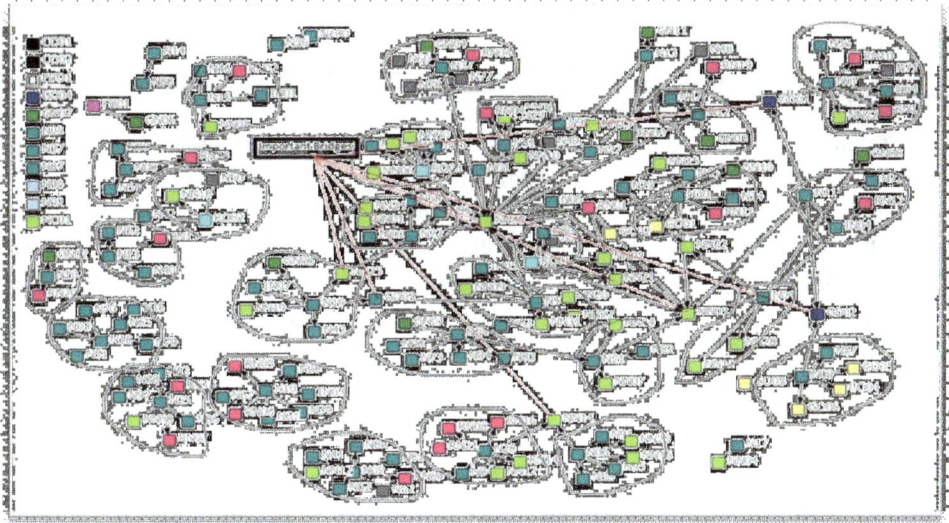

WNTA (NN33)

Action Aid (GD01)

Save the Children Foundation (GD02)

UNICEF (GD03)

VANI (NN03)

FORCE (NN14)

Nine is Mine (NN25)

The importance of the bridging organizations is that they are playing a crucial role in connecting other network sub groups. Suppose if it is required share any information among CSOs under WNTA network. Then definitely we need to be in touch with the bridging organizations mentioned above. In the above diagram we can see that some of the CSOs are isolated and not connected with any big network. Referring to the above diagram this can be stated that adopting this approach may be useful for a programme planner about how to reach such isolated CSOs.

Discussion and Conclusion

Social network analysis and its pictorial presentation gave us an opportunity to map networking pattern of CSOs. From this pictorial presentation this becomes easy to understand the strength of the network and to find out key network members without doing any complex mathematical analysis.

This paper intended to share process oriented learning while conducting a social network analysis within the purview of Oxfam India's work with partner CSO networks. Outcome of the study has witnessed uunderstanding of network structure and multi directional relationship of Civil Society Organizations (CSOs) in India working on issues of Inequality. Indian CSOs have witnessed formation of partnership and collaboration based on ideological similarities. NGOs/CSOs working on to address various social issues in India have deliberated on potential use of several advocacy tools and initiatives. Although 'network structure' of CSOs in India is not properly institutionalized but some of the National and International Networks of NGOs and CSOs indeed made credible beginnings.

"Connectedness" is the key element in social networks. To be part of a social network, each member must have either actual or potential links to at least one other member of the network. These links may be direct or indirect. For example, two CSOs can be interlinked through resource sharing or advocacy. While some members may be peripheral in the network or almost completely isolated, each one must somehow be connected to other members if it is to be considered part of the network. This approach helped us to visualize the visual pattern of network function and multi-directional relationship among Indian CSOs those are constructed within national network. This attempt also gave us an overall picture how CSOs within a network are associated with other multiple networks. For a whole network analysis within ambit of Oxfam's collaboration with other key networks, visibility on CSOs' networking pattern with one another found prominent but it is difficult to understand and explain well defined structure and strength of the networks. Higher rate of participation through online e-survey using survey monkey link appeared a bit hurdle. CSOs' capacity to understand use of online data filling remained weak link.

Network analysis is an analytical tool to assist in understanding and help in decision-making. It is becoming increasingly popular as methodology for understanding complex patterns of interaction among networks. The network perspective examines actors (CSOs) that are connected directly or indirectly by one or more different multidirectional relationships. Regardless of unit level, network analysis describes structure and patterns of relationships, and seeks to understand both their causes and consequences. Understanding of network structure and relationship is necessary for any collaborative work. For example, if we wish to know who are the key CSOs in India are working on certain issues (*e.g.,* Health) and how these organisations are inter-related to each other through different kinds of relation, network analysis will be helpful to furnish, 'connectedness and types of relationship' among CSOs for using knowledge, advocacy and policy planning.

Acknowledgement

The opinions expressed in this paper are those of the author(s) and do not necessarily reflect those of Oxfam. The paper is based on meta-reviews and baseline surveys conducted by Oxfam India during implementation of European Commission-NSA supported project. The aim of the EC-NSA supported project is focused on "Empowering Civil Society Networks in an Unequal and Multi- Polar World". Some of the data represented here do not necessarily reflect views of Oxfam and or partner organizations like Wada Na Todo (WNTA). Facts are established based on voluntary participation and non-judgmental responses collected through primary and secondary sources from Civil Society Network organizations. Sincere thanks and appreciation goes to the partner organization networks – PBI and WNTA, all CSO partner organizations who have had participated and shared their consent. We acknowledge European Commission and Oxfam Global Programme Unit, Oxfam GB and Oxfam India – India and the World team for sharing feedback at every step involved. A special thanks to the communities who shared information and participated whole heartedly in the process of baseline assessment.

References

Burt,R.S., Minor,M.J., and Associates.(1983). *Applied network analysis*. Beverly Hills, CA: Sage.

Dershem, L., T. Dagargulia, L. Saganelidze, S. Roels. (2011). NGO Network Analysis Handbook: how to measure and map linkages between NGOs. Save the Children. Tbilisi, Georgia.

Dutta,B., and Jackson,M.(Eds.).(2003). *Networks and groups: Models of strategic formation*. Berlin: Springer Verlag.

Wassermann, S., and Faust, K. (1994). *Social network analysis: Methods and applications*. Cambridge, UK: Cambridge University Press.

http://csnbricsam.org/documents/

Chapter 21

Innovation in Evaluation to Inform Policy Convergence: Complex Systems Approach to Assess Entrepreneurship-driven Intervention in eKutir's VeggieKart

Spencer Moore[1]*, Srivardhini K. Jha[2,3], Suvankar Mishra[4], Daniel Ross[5], Summer Allen[3] and Laurette Dubé[2,6]

[1]*Arnold School of Public Health, Department of Health Promotion, Education, and Behavior, University of South Carolina, USA*
[2]*McGill Centre for the Convergence of Health and Economics (MCCHE), McGill University, Canada*
[3]*International Food Policy Research Institute (IFPRI), Washington, D.C., USA*
[4]*EKutir, Bhubaneswar, Odisha, India*
[5]*Wholesome Wave, Bridgeport, CT, USA*
[6]*Desautels Faculty of Management, McGill University, Canada*

ABSTRACT

Convergent Innovationhas been advanced as a solution-oriented paradigm to address the complex human problems at the nexus of agriculture, food and health by defining new paths of convergence between economic growth and human development.Innovative approaches to policy convergence require innovative evaluation strategies and methods. The following study provides a description of the evaluation strategies used to assess the effectiveness of the VeggieKart intervention in Odisha, India. EKutir established VeggieKart as a farmers' market for fresh fruits and vegetables in low-income communities in Odisha,

* *Corresponding Author:* E-mail: mooreds4@mailbox.sc.edu
Address: 915 Greene Street, Room 529, University of South Carolina, Columbia, SC, 29208.

India. TheVeggieKart intervention consists of a pilot study of a convergent innovation approach to increase agricultural production, nutritional intake, and health in resource poor groups of rural and urban communities, with a focus on women and children. The VeggieKart intervention was designed to be a nutrition-sensitive, multi-agent and –site intervention targeting systems-level change in the F2C value chain with concomitant impacts on household vegetable consumption. To assess the effects of the VeggieKart intervention, a mixed methods approach (survey methods, social network analysis, qualitative interviews) was applied at multiple ecological levels – individual, interpersonal, organizational, community and systems. Intervention findings will contribute to recommendations for policy convergence across different governmental sectors in India.

Keywords*: Convergent innovation, Evaluation, Complex systems, Healthy eating.*

Introduction

The agriculture-food value chain is at the core of economic growth and human development in India. Agriculture and food sectors contribute over 20 per cent to India's GDP and employ more than 50 per cent of India's population. At the same time, the agri-food sectors also shoulder the onus of securing the nutritional security for a population that is simultaneously fighting undernutrition and overnutrition. These challenges persist in spite of billions of dollars being allocated by the Indian government and international organizations to alleviate rural poverty and to secure mother/child nutrition (The Hindu News, 2013).

Given the inter-related nature of agriculture, food and health (Hammond and Dube 2012), there is an urgent need to build policy synergiesamong these three sectors to drive transformation at scale. Yet, building such synergiesis complicated and the present approach to policy coherence is insufficient (Dube *et al.,* 2014a; Dube *et al.,* 2014a; Pingali 2012). The Whole-of-Society (WoS) approach to policy *convergence* weaves economic and health sector considerations together, while placing health at the core of innovation, strategy and operation in mainstream economic sectors, including agriculture and food (Dubé *et al.,* 2012; Dube *et al.,* 2014b; Arora *et al.,* 2014; Aady *et al.,* 2014). Convergent Innovation (CI) is a form of *meta-innovation*—an innovation in the way that we innovate (Dubé *et al.,* 2014c; Jha *et al.,* 2014). CI has been advanced as a solution-oriented paradigm to go beyond what has been possible thus far in addressing complex human problems at the nexus of agriculture, food and health by defining new paths of convergence between economic growth and human development. Figure 21.1 illustrates the CI architecture. CI is hinged on four enabling conditions – in-depth knowledge of human behavior and decision-making, strategic engagement by private enterprises, cross-sectoral collaboration and a robust digital infrastructure. Building on these enablers, CI bundles technological innovations in agriculture, food, nutrition and health, with social and institutional innovations to scale up and accelerate the transformation of rural and urban communities toward a convergence of their human and economic development. This type of innovation entails changes in the way that individuals, community organizations, and value chain actors interact with each other, thereby creating additional economic opportunities. Micro-entrepreneurial activities, for example, can

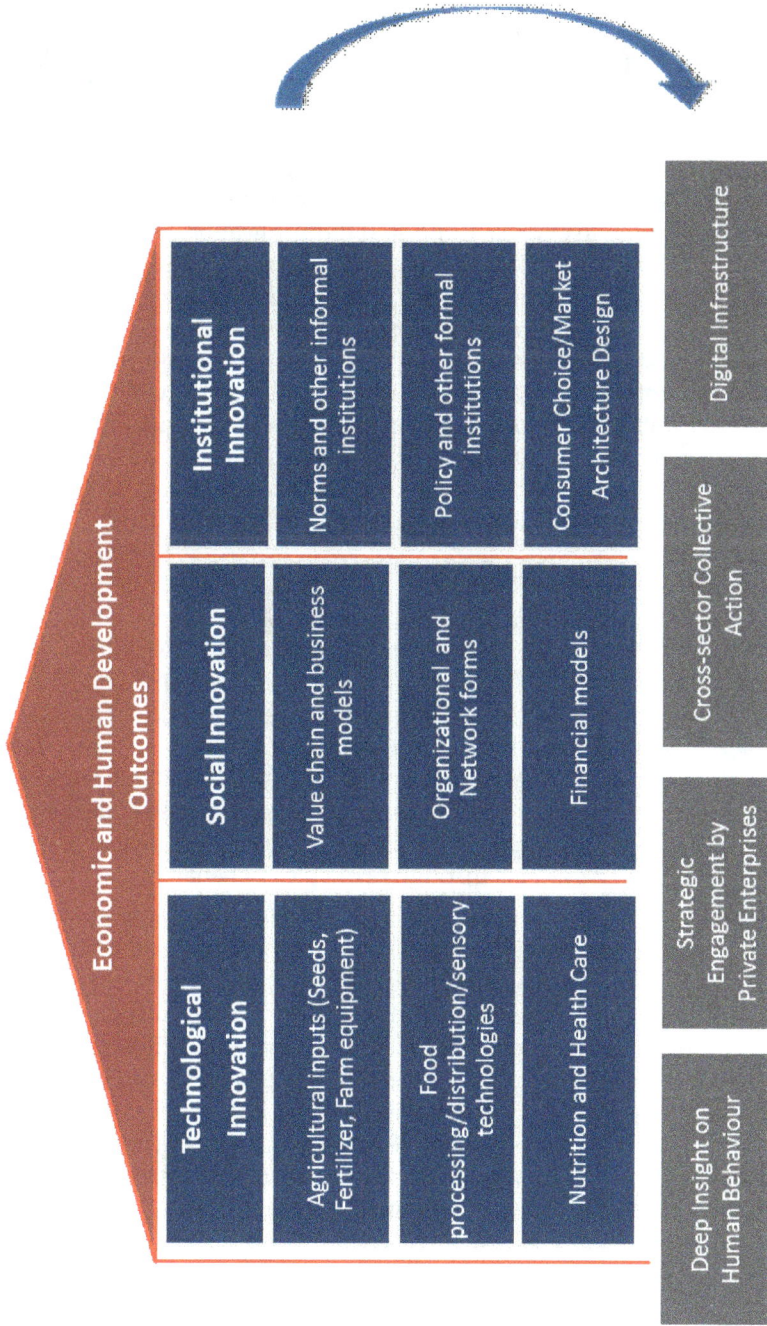

Figure 21.1: Conceptual Model of Convergent Innovation.

help catalyze local innovation and self-reliance, enabling communities in underserved populations to provide goods or services locally and sustainably through financial or non-financial exchange. CI advances a new way of tackling complex problems but the operationalization of CI is dependent on the focal sector, the activity (the product/ initiative) in question and the scale of such activities.

The following paper discusses CI approaches to human and economic development in the context of VeggieKart, an intervention based on micro-entrepreneurial principles in Odisha, India, and the application of complex-systems principles to the evaluation of CI interventions. CI ultimately involves intervening into a complex system of economic, social and institutional relationships. Complex systems have a number of defining characteristics, including the possibility of non-linear change, the existence of emergent properties, and the importance of feedback mechanisms in the unfolding of an intervention (Shiell, Hawe, Gold, 2008). These characteristics have implications for evaluation strategy and design. Complex systems approaches recognize the fact that new structures, institutions, and relationships may emerge from intervention activities, requiring mixed-method designs and the measurement of intervention outcomes at multiple levels.

eKutir and the VeggieKartValue Chain

EKutir is a social enterprise that leverages local micro-entrepreneurial motivations and processes to solve smallholder farmers' poverty through a distribution network of digitally-trained entrepreneurs, market linkages, technology, and data. The model serves as an engine for the creation of micro-entrepreneurs in a diversity of sectors, planting the seed for convergent innovation among rural villages. EKutir uses this model to transform agriculture and nutrition linkages in rural communities, and between these areas and slum and other poor urban communities in the state of Odisha, India. EKutir builds capacity and fosters linkages between vulnerable populations and local food systems through principles of behavioral economics.

EKutir established VeggieKart as a farmers' market for fresh fruits and vegetables in low-income communities in Odisha, India. The traditional vegetable supply chain from producer to end consumer is disaggregated, with a long series of middlemen between the smallholder farmer and the end consumer. This chain has been shown to be ineffective, unreliable, and result in considerable wastage (25-30 percent), with low price realization for the farmers and lack of fresh, quality vegetables for the consumers (Narrod, *et al.,* 2009). The VeggieKart value chain reduces intermediaries through the introduction of micro-entrepreneurial structures and a novel distribution channel, bypassing the Mandi market system and the brokers associated with these markets, reducing losses due to poor handling, fees, rents, interest, and even graft.

Figure 21.2 illustrates the F2C VeggieKart value chain model. The VeggieKart value chain consists of several actors: farmers, agricultural micro-entrepreneurs (agri-entrepreneurs), the VeggieKart distribution channel, vegetable micro-entrepreneurs (veggie-entrepreneurs), and consumers.

Agricultural Micro-entrepreneurs

Agri-entrepreneurs provide inputs, technical assistance, and market linkages to smallholder farmers who have chosen to participate in eKutir agricultural programming. The Veggie Kart "Agri-Entrepreneurs" provide inputs, technical assistance, market linkages, and helpful demand information so the smallholder farmer can better plan their production and harvest, reducing risk and increasing value. To facilitate localized interactions with farmers, agri-entrepreneurs organize eKutir farmers into groups of 15-25 members called Farmer Intervention Groups, with possibly multiple groups per village. As part of their activities, agri-entrepreneurs aggregate the produce cultivated by these groupsat local aggregation points for insertion into VeggieKart distribution channels.

The VeggieKart Distribution Channel

VeggieKartpurchases vegetables from eKutir farmers at local aggregation points and transports them to a central warehouse, where quality control checks are undertaken. After weighing, sorting, grading, packaging, and re-weighing the vegetables, the produce is sent to the farmers' marketsestablished in different parts of the capital city, Bhubaneswar. The distribution channel of VeggieKart consists of the following structures:

1. Online: eCommerce platform and direct doorstep delivery
2. VeggieMart: Converting existing pawnshops and mom-and-pop shops into farmers' market, branded by VeggieKart.
3. VeggieWheels: Converting existing vegetable vendors into vegetable entrepreneurs with push carts across thoroughfares in the city.
4. VeggieLite: Establish vegetable entrepreneurs in low-income communities to sell vegetables at reduced prices.

Through the diverse range of VeggieKart farmer markets, farm-fresh food is made accessible to all types of consumers at affordable prices.

Vegetable Micro-entrepreneurs

Vegetable micro-entrepreneurs are the local vegetable vendors who have been converted into an organized chain of retailers, earning a livelihood by selling the vegetables sourced by the different VeggieKart distribution channels. These entrepreneurs operate in small stores (Veggie Mart) or are stationed with push carts on thorough fares (Veggie Wheel) within Bhubaneswar. VeggieLite entrepreneurs operate in rural villages that are part of eKutir programming and urban areas to distribute and sell the vegetables to low-income, resource-poor communities. The vegetables are distributed through delivery pointsestablished in thoroughfares of the capital city, local slums, and in rural clusters of villages.

The VeggieKart Pilot Intervention

Funded through the the Grand Challenge India Funding Opportunity "Achieving Healthy Growth Through Agriculture and Nutrition," the VeggieKart project consists of a pilot study of an innovative entrepreneurial approach to increase agricultural

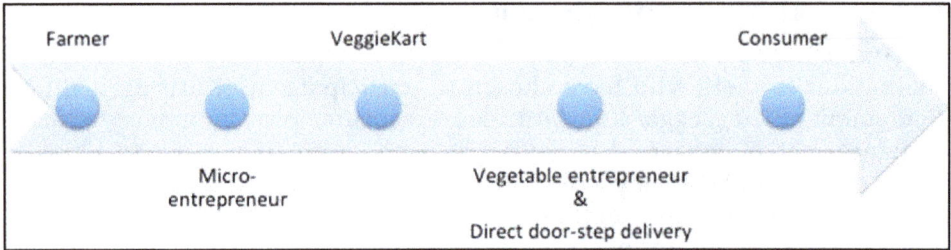

Figure 21.2: The VeggieKart Vegetable Value Chain.

production, nutritional intake, and health in resource poor groups of rural and urban communities, with a focus on women and children. The VeggieKart intervention was designed to be a nutrition-sensitive, multi-agent and –site intervention targeting systems-level change in the F2C value chain with concomitant impacts on household vegetable consumption. The project weaves together the goals of agricultural, nutritional, and social innovation. Agricultural innovation involves the promotion of good agricultural practices in growing vegetables and facilitating this through agri-entrepreneurs. Nutritional innovation involves the identification of vegetables that are highly nutritious and rich in iron and mineral content so that smallholder farmers gain access to due information and knowledge on production of such vegetable crops. Social innovation also occurs on multiple levels and involves the identification of the community structures and organizational features that lead to greater efficiencies in the vegetable value chain and policy convergence.

The pilot study is meant to test primarily the effects that veggie-entrepreneurs catering to low-income rural and urban communities have on household vegetable consumption, and (2) explore whether governmental nutritional schemes can be leveraged to increase the affordability of vegetables to low-income consumers. For the first objective, the VeggieKart intervention will increase the availability and accessibility of quality vegetables by increasing the number of veggie-entrepreneurs and establishing new distribution centers in rural villages and urban slums. Current VeggieKart efforts will be leveraged to increase the number of vegetable entrepreneurs from 23 to 80 through the intervention, with 40 percent of the new entrepreneurs being women. To accommodate this increase in distribution points, the intervention also entails on the production side raising the farmer base to 1,000 and the number of agri-entrepreneurs, and incorporating gender inclusivity into farming practices. Women will be empowered and benefitted in multiple roles as smallholder producers, vegetable-entrepreneurs, and as consumers of fresh produce for their households, thereby creating economic opportunities for women as producers and entrepreneurs (sellers of vegetables) throughout the value chain.

The second objective of the pilot intervention is to develop a behavioral incentive intervention to improve the appeal of fresh fruits and vegetables (vis-à-vis less nutritious alternative) through the use of governmental financial support provided to the poor, and to provide preliminary evidence on whether such an incentive might impact consumer purchasing behavior at VeggieKart markets. This sub-intervention will be conducted in one rural and one urban location.

Evaluating a CI Intervention

To evaluate the VeggieKart pilot intervention, a mixed methods approach was applied at multiple ecological levels – individual, interpersonal, organizational, community, and systems – and in rural and urban settings. Although the benefits of the intervention are meant to accrue at the individual level in the form of increased household vegetable consumption, intervention consequences may ensue at and across different ecological levels. Figure 21.3 provides a schematic of the CI VeggieKart evaluation strategy.

Individual Behavioral Change

To assess the individual-level effects of the VeggieKart intervention on vegetable consumption, the evaluation employs a quasi-experimental design with pre- and post-treatment measures in rural and urban settings. Using this design, two separate study samples will be recruited: (1) a rural sample of producers/consumers, *i.e.*, farmers and their household members, and (2) an urban sample of consumers. In rural areas, the quasi-experimental design will feature three groups: (1) eKutir farmer households in villages where eKutir delivers services; (2) non-eKutir farmer households in eKutir villages; and (3) non-eKutir farmers in non-eKutir villages. Villages from which these groups are drawn will be matched on salient village characteristics(*e.g.*, farmer density, clear need, the presence of non-governmental organizations). In urban areas, the quasi-experimental design will feature two groups: (1) a sample of urban consumers residing in a Veggie-Kart catchment area and (2) another sampleof urban consumers who do not reside in an area serviced by Veggie-Kart.

Rural and urban study participants will be administered a household questionnaire consisting of a number of modules: (1) household geographic and demographic characteristics; (2) fruit and vegetable consumption, nutritional behavior, and food security; (3) agricultural practices (rural households); (4) social capital and social networks; and (5) household income and poverty level. To assess fruit and vegetable consumption, participants will be presented an inventory of culturally-and geographically-relevant fruits and vegetables and asked whether they had consumed each vegetable over the past seven days. If so, follow-up questions asked participants to describe the amount, price, and place the vegetable or fruit was purchased and how much may have come from self-production. These items allow assessment of total volume and diversity of fruit and vegetable consumption.

Interpersonal Relations and Structures

To assess interpersonal relationships and structures, study participants were administered a name generator/interpreterinstrument as part of the household questionnaire. The name generator questions ask participants to name up to five people with whom they have spoken about (1) farming and agricultural matters and (2) dietary matters over the past three months. The ensuing name interpreter asks participants questions about those people with whom they have spoken (*e.g.*, village of residence, occupation). Name generator and interpreter items will allow the calculation of a range of social network measures, including network size, village

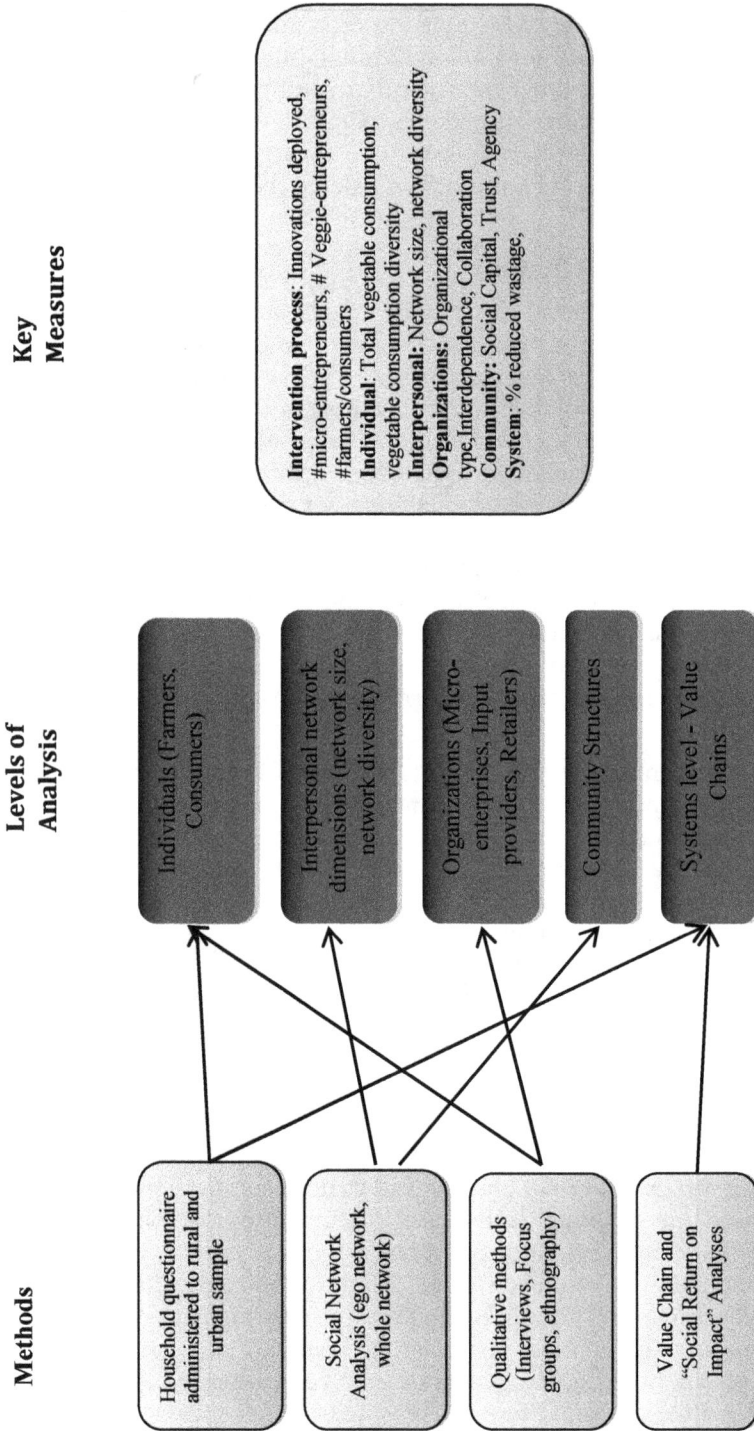

Key Measures

Intervention process: Innovations deployed, #micro-entrepreneurs, # Veggie-entrepreneurs, #farmers/consumers
Individual: Total vegetable consumption, vegetable consumption diversity
Interpersonal: Network size, network diversity
Organizations: Organizational type, Interdependence, Collaboration
Community: Social Capital, Trust, Agency
System: % reduced wastage,

Levels of Analysis

- Individuals (Farmers, Consumers)
- Interpersonal network dimensions (network size, network diversity)
- Organizations (Micro-enterprises, Input providers, Retailers)
- Community Structures
- Systems level – Value Chains

Methods

- Household questionnaire administered to rural and urban sample
- Social Network Analysis (ego network, whole network)
- Qualitative methods (Interviews, Focus groups, ethnography)
- Value Chain and "Social Return on Impact" Analyses

Figure 21.3: The Convergent Innovation Evaluation Strategy: VeggieKar

diversity, and occupational diversity. These measures will be used to assess whether interpersonal network characteristics affect vegetable consumption behavior and whether there are changes in interpersonal network characteristics over the VeggieKart intervention period.

Organizational and Inter-organizational Levels

The VeggieKart ecosystem comprises several small and large organizational actors – micro-entrepreneurs (on supply and distribution end), agricultural input providers (seed, fertilizer companies), and retail outlets - who need to act in coordination in order to drive their own economic growth and the human development of the larger population. Such collaborative action requires goal alignment, and supporting structures and processes to enable information exchange and coordination. Data on the emergence of these elements will be gathered using semi-structured interviews and whole social network analysis. The qualitative data from interviews will address the 'how' questions, isolating the underlying mechanisms that lead to inter-organizational coordination. Second, organizational representatives will be administered a questionnaire with a network module asking about their resource exchanges with other organizations. These data will be used to measure the inter-organizational dynamics and map the structure of integrative collaboration that unfolds among the 25-30 organizational actors over the course of the intervention.

Community Structures

To assess community features and structures, the intervention and evaluation will focus on measuring social capital as an intervention facilitator and outcome (Moore, Salsberg, Leroux, 2013). Social capital refers to the resources available to individuals and groups through their social networks. Social capital will be measured at the individual level using items from the social capital and network module of the household questionnaire and aggregated to the village or neighborhood level to capture community social capital (Moore, Buckeridge, Dubé, 2014).

Systems Level Transformations

Systems-level transformations will be evaluated through an analysis of the VeggieKart value chain and in terms of an integrative metrics of "social return on impact." First, value chain analyses (VCA) will be informed by data gathered from the household surveys of smallholder producers and consumers and the tracking of 3-4 specific products through traditional and Veggie Kart value chains as per the stacked survey approach that uses surveys in each segments of the value chain, including wholesalers and retailers (Reardon, *et al.,* 2014; Joshi 2014). Second, the financial sustainability of Veggie-Kart intervention will be examined with an adaptation of the metrics of financial sustainability developed by Wholesome Wave for analysis of food hubs in the USA.

Next Steps and Contribution

Findings from the intervention will have important implications for how development activities can jointly optimize on economic and human development outcomes. The complex system evaluation strategy promises to provide a rich understanding of ecosystem dynamics. Findings will inform the implementation of coordinated efforts and policy coherence among governmental departments and ministries to foster community mobilization and value chain innovation for nutritious agricultural commodities and for reaching the scale of transformation needed to curb food security and nutrition deficiency at a population level.

Acknowledgements

The VeggieKart intervention is funded in part through an award from the first Grand Challenge India Funding Opportunity "Achieving Healthy Growth Through Agriculture and Nutrition" with support from the Department of Biotechnology (DBT), Ministry of Science and Technology, Government of India and Bill and Melinda Gates Foundation (BMGF). Support for the research is also provided by a convergent innovation partnership development grant for the Social Sciences and Humanities Research Council of Canada and by CGIAR-Agriculture for Nutrition and Health.

References

Addy, NA, Poirier, A,Blouin C, Drager, N, and Dubé, L. 2014. WholeofSociety Approach for Public Health Policymaking: a Case Study of Polycentric Governance from Quebec, Canada. *Annals of the New York Academy of Sciences*, 1331: 216-229.

Arora, NK., Pillai R, Dasgupta R, and Rani Garg, P. 2014. Whole-of-society monitoring framework for sugar, salt and fat consumption and communicable diseases. *Annals of the New York Academy of Sciences*, 1331: 157-173.

Dubé, L, Webb P, Arora N, Pingali P. 2014a. Agriculture, Health, and Wealth Cconvergence: Bridging Traditional Food Systems and Modern Agribusiness Solutions.*Annals of the New York Academy of Sciences*, 1331: 1-14.

Dubé, L, Addy NA, Blouin, C, and Drager, N. 2014b. From Policy Coherence to 21st Century Convergence: A Whole of Society Paradigm of Human and Economic Development. *Annals of the New York Academy of Sciences*, 1331: 201-215.

Dubé, L, Jha SK, Faber, A, Struben J, London T, Mohapatra A, Drager N, Lannon C, Joshi PK and McDermott, J. 2014c. Convergent Innovation for Sustainable Economic Growth and Affordable Universal Health Care: Innovating the Way We Innovate. *Annals of the New York Academy of Sciences*, 1331: 119-141.

Dubé, L, Pingali, P, Webb, P. 2012. Paths of Convergence for Agriculture, Health, and Wealth. *Proceedings of the National Academy of Sciences,*109(31): 12294-12301.

Hammond, RA,Dubé, L. 2012. A Systems Science Perspective and Transdisciplinary Models for Food and Nutrition Security. *Proceedings of the National Academy of Sciences*, 109 (31): 12356-12363.

Jha, SK, McDermott, J, Bacon, G,Lannon, C, Joshi, PK, Dubé, L. 2014. Convergent Innovation for Affordable Nutrition, Health, and Health Care: The Global Pulse Roadmap. *Annals of the New York Academy of Sciences*, 1331: 142-156.

Joshi, P.K. 2014. Dynamics of Pulses in Indian food system: technologies, policies and institutions from farm-to-fork.IUFoST, 17th World Congress of Food Science and Technology, Montreal.

Moore, S, Buckeridge, D, Dubé, L. 2014. Advancing social capital interventions from a network and Population Health Perspective in Global Perspectives on Social Capital and Health. Kawachi, I, Takao, S, Subramanian SV (eds). New York: Springer Press.

Moore, S, Salberg, J, Leroux J. 2013. Advancing social capital interventions from a network and Population Health Perspective in Global Perspectives on Social Capital and Health. Kawachi, I, Takao, S, Subramanian SV (eds). New York: Springer Press.

Narrod, C, Roy, D,Okello, J,Avendano, B, Rich, K, Thorat, A. 2009. Public-private partnership and collective action in high value fruit and vegetable supply chain. *Food Policy*, 34: 8-15.

Pingali, P. 2012. Green Revolution: Impacts, limits, and the path ahead. Proceedings of the National Academy of Sciences USA, 109(31): 12302-12308.

Reardon, T, Chen, KZ,Minsten, B, Adriano, L, Dao, TA, Wang, J, Das Gupta, S. 2014. The quiet revolution in Asia's rice value chains. *Annals of the New York Academy of Sciences*, 1331: 106-118.

Struben, J, Chan, D,Dubé, L. 2014. Policy Insights from the Nutritional Food Market Transformation Model: The Case of Obesity Prevention. *Annals of the New York Academy of Sciences*, 1331: 57-75.

Shiell A, Hawe, P, Gold, L. 2008. Complex interventions or complex systems? Implications for health economic evaluation.*BMJ*, 336 (7656): 1281-1283.

The Hindu Website: World Bank promises big push to poverty alleviation schemes in India. 2013. *The Hindu News*. March 14th. Available online at: http://www.thehindu.com/news/national/world-bank-promises-big-push-to-poverty-alleviation-schemes-in-india/article4506331.ece

Chapter 22

Innovation in Evaluating Humanitarian Response: Post-tsunami Lessons from Sri Lanka

Cynthia Caron

Assistant Professor, International Development and Social Change,
Clark University, MA, USA
E-mail: ccaron@clarku.edu

ABSTRACT

The death and destruction following the 2004 Indian Ocean or Boxing Day Tsunami is well documented including the displacement of over a million persons in Sri Lanka alone. Governments and humanitarian actors have multiple information needs to facilitate and ultimately evaluate relief, recovery and reconstruction processes following such a complex emergency. Given the scale and geographical scope of displacement in Sri Lanka and the sheer number of organizations that entered the country to help save and rebuild lives, the relief and recovery effort was initially uncoordinated.In order to stem confusion and harmonize the delivery of relief, the UN spearheaded a joint monitoring and evaluation exercise or joint mechanism. In this chapter documents the experience of establishing this joint mechanism to coordinate and deliver up-to-date information on tsunami relief and recovery efforts tothe Government of Sri Lanka and humanitarian actors, particularly with respect to the construction and maintenance of transitional shelter sites housing tsunami-displaced families. Ithighlights innovations to the mechanism's process and tools over the project cycle and some of the implementation challenges faced. The chapterends with a set of recommendations for supporting longitudinal assessments to evaluate and track the effectiveness of humanitarian response in complex emergencies.

Keywords: *Post-disaster management, Joint mechanism, Personal digital assistant (PDA), Sri Lanka.*

Introduction

The death and destruction following the 2004 Indian Ocean or Boxing Day Tsunami is well documented. In Sri Lanka alone, the tsunami's waves killed over 35,000 persons and displaced overa million more (Government of Sri Lanka and Development Partners, 2005). At the end of 2005, the Government of Sri Lanka (GoSL) along with donors and humanitarian partners had established more than 400 shelter sitesacross 11 districtsand constructed over 53,000 individual transitional shelters to house the displaced until they could be resettled (ibid).After the quick and rather uncoordinated construction of shelter sites, there was a general confusion among agencies as well as the GoSLabout where all the sites were. In mid-2005, the Humanitarian Information Centre (HIC), the Office for Coordinating Humanitarian Assistance (OCHA), and the Government of Sri Lanka's Transitional Accommodation Project (TAP) conducted an assessment of all transitional shelter sites to locate and map them, determine what infrastructure work remained to be done and maintained to ensure that people were living in dignity, as they waited for new permanent homes.

Following a consultative process at the district level with local government officials, I/NGOs and UN agencies that were responsible for building and maintaining shelter sites designed a common assessment form. First, agency and government representatives agreed upon basic living standards in shelter sites, many of which aligned with the internationally recognized SPHERE standards. Second, each sector lead agency includedassessment questions. WHO and UNICEFcreatedquestions for the health section, UNICEF provided feedback on questions designed for the education, water and sanitation sections and together with UNHCR suggested questions on child protection and gender-based violence. ILO and FAO design questions on livelihoods and resettlement. All agency representatives involved in the process collectively decided what types of demographic information was necessary for useful data disaggregation. This joint mechanismbecame the Transitional Shelter Site Tracking (TSST) Project.

The Transitional Shelter Site Tracking (TSST) Project as Joint Mechanism

HIC and OCHA funded the first the Transitional Shelter Site Tracking (TSST) assessment in 2005, withUNICEF providing funding in 2006 and 2007. The International Organization for Migration (IOM) and the American Red Cross funded a separate assessmentin 2007. The United Nations Office for ProjectServices (UNOPS) implemented all assessments.[1] The TSSTproject was the first joint assessment undertaken in Sri Lanka, with UN agencies and large INGOsworkingtogether with the GoSL's post-tsunami administrative system to develop a multi-cluster assessment tool for post-disaster assessment and relief operation monitoring. By using one tool and placing responsibility for assessments with one agency (UNOPS), assessment exercises were neither replicated nor duplicated, saving both time and money, while alsoreducing research fatigue for respondents and creating a consolidated data management system.

---

1   The TSST project ended in December 2007 not because the transitional sites were all closed, but due to lack of funding. Many donors committed to only three years of post-tsunami programming.

TSST Methodology and Method

Staffing and Training

Every six months, the TSST enumeration team interviewed shelter site managers and walked each site, measuring length, width and height of shelters, inspecting toilets to make sure they were cleanand well lit,observing if the site was well-drained and free of garbage, and so on. The assessment team collected gender-disaggregated information across all sectors (*i.e.*, # of female headed-households, # of toilets for men and women, reported incidents of gender-based violence).The team included a project manager, a technical coordinator, computer programmers (with knowledge of Geographic Information Science or GIS), a quality control officer, and male and female enumerators. Mixed gender teams of two individuals (male and female) were deployed to each siteto collect gender-disaggregated data in a culturally appropriate way.

Staff training included an introduction to TSST objectives, the assessment tool, SHERE standards (including how to use a tape to properly measure shelter dimensions), the useof hand-held computers or Personal Digital Assistants (PDAs). To facilitate data collection and analysis, EmergencyInfo,a software program developed by DevInfo was installed.The jointly created assessment tool(question and answer choices)wasprogrammed inEmergencyInfo and uploaded onto the PDA. Enumerators recordedand saved answers to the PDA.[2]Immediately following a site visit the enumerator emailed the data file to the project office for processing, quality control, and reporting. Assessment results often were available on the same day as the site visit.

Project Outputs (reporting)

TheTSST teamgenerated standardized reporting templates and createdGIS maps to display results visually. Reports and maps were posted on the project-dedicated website, www.unops.org.lk/tsst. For humanitarian partners in remote areas with intermittent Internet service, the project manager would fax specific reports on request.The team createdand posteda wide range of outputsfor free including:

1. Individual site-level reports
2. Decommissioned site reports, when applicable
3. District level vulnerability profiles
4. District level summary reports
5. Raw data tables for agency staff to undertake their own analyses
6. GIS maps to visually displaydata.[3]

-----

2 For more on EmergencyInfo and the TSST project, see DevInfo. *Tracking Post Tsunami Reconstruction Effort in Sri Lanka.* August 2007; http://www.devinfo.org/libraries/aspx/diorg/pdfs/GPS_Geo-spatial_r3.pdf; accessed on 27 December 2014.

3 For an example of a TSST results map, see http://reliefweb.int/map/sri-lanka/sri-lanka-galle-matara-hambantota-ampara-and-trincomalee-districts-population-trend; accessed on 26 December 2014.

As a humanitarian response project, TSST project staff worked closely with partners and updated the process to provide relevant information to ensure hygienic living conditions and facilitate resettlement. The implementation of a joint mechanism is itself an innovation to evaluation implementation. Next I discuss briefly three specific project innovations before examining a few implementation challenges.

Innovation 1: Responsive Tool Development

The initial assessment form evolved, as agencies moved from relief to resettlement and livelihood restoration. The original form included 75 questions in 9 sections. The 2007 form included 141 questions and 17 sections. The 2007 form included expanded sections for shelter, water and sanitation, health, education, risks to vulnerable groups and new sections on livelihoods, food and nutrition, and community-based site management

As families moved out of transitional shelter sites, site managers requested a site-decommissioning checklist. The checklist ensured that all pit latrines were emptied and filled in and that all solid waste and shelter materials including cement foundations were removed. The site decommissioning process aimed to return the site to the landowner (government or private) in a condition as close as possible to its original state. As the resettlement process accelerated, shelter sites closed as families moved out of them and into permanent houses. The monitoring data showed that the resettlement process appeared to be moving more quickly in some districts than in others, prompting humanitarian actors to ask new questions about why some families were left behind.

Innovation 2: Mixing Methods in the Post-disaster Evaluation Context

As resettlement progressed, the data indicated that the number of camps and the number of families residing in them was decreasing. However, in the Ampara (See footnote 3) and Colombo districts, thousands of families remained in transitional shelter nearly 3 years after the tsunami hit (Caron, 2009). Humanitarian actors needed to understand why this was the case. In late 2007, the Centre on Housing Rights and Evictions (COHRE) funded a mixed methods study that added a series of open-ended questions that allowed agencies to learn, from the perspective of shelter site residents themselves, how they understood the resettlement process to work and documented their interactions with government and NGO officials, as they tried to access resources to move into permanent homes. The additional qualitative component asked 'how' and 'why' questions that focused on the *process* of trying to leave shelter sites and led to the creation of legal aid clinics to help residents' access and use resettlement assistance; ultimately improving the delivery of resettlement services (ibid).

Innovation 3: Developing a Site-level Vulnerability Indicator

With some tsunami-affected districts having more than 200 individual shelter sites, humanitarian agencies asked for a *single number* to help them quickly and comparatively assess which shelter site residents might be most vulnerable to risks and hazards. Using the answers collected on the common form, 12vulnerability indicators were constructed to calculate a total site-level vulnerability indicator (UNICEF and UNOPS, 2006; See Appendix 1).

The development of vulnerability indicators followed the samecollaborative and participatory process with Government and I/NGOsrepresentatives.For illustrative purposes, I explain the construction of the healthcare (HC) vulnerability indicator here. Once agency representatives agreedupon seven variables for its calculation (UNICEF and UNOPS, 2006: 15-16), vulnerability indicator calculation started with *recoding* the data on a weighed 1 to 4 sliding scale (See Appendix 2).Next,all seven variables were not considered to be of equal importance. To indicate the more important variables the *recoded score*was multiplied (weighed) by an agreed-upon factor. For example, for the HC indicator, variable HC2 'distance to emergency obstetrics' was considered the most importantvulnerability variable, with the recoded score for 'distance to emergency obstetrics'multiplied by '6' (a first aid box on site was less important and multiplied by'1').After recoding and multiplying by the weighed value, the new values were added together and divided by the sum of the weighs. In the case of the HC indicator, the sum of the weightswas 26, for a final HC indicator calculation of:

$$\frac{HC1+HC2+HC3+HC4+HC5+HC6+HC7}{26} = \text{numerical indicator for HC vulnerability ranking}$$

The higher the final number calculated, the more vulnerable or less satisfactory were a site's access to health care services.While agency representatives and government officials found a singular indicatoruseful for comparative purposes, there was general consensus thatindicator construction could have been more meaningful(Meeting notes, March 2007; see further discussion below).

Constraints and Action Taken

While flexibility, adaptability, and willingness of all partners to collaborate enabledthe success of TSST as a joint M and E exercise, the project did face a number of constraints, which I address briefly below.

Coordination between Agency Representatives

TheUN sector or cluster approach to the TSST assessment was a shortcoming. Working jointly was a new process, with agencies organized aroundsingle-issue clusters such as water, shelter, or sanitation, which precluded examining connections and synergies across issues. As the variables in Appendix 2 show, the HC indicator is a measure of*accessibility*to healthcare services, based on the assumption that reduced access increases vulnerability. It did notuse physical site datacollected under other 'clusters' sections of the common assessment form.Common sense says that good drainage and waste removal are necessary to keep sites free of mosquito-breeding areas(dengue prevention), unfortunately data collected under flooding or waste management was not used for HC vulnerability indicator creation. Data is notnecessarily discrete. The construction of the HC indicatoris a particularly illustrative example of the limitations involved whena single cluster ideology prevails over and precludes possible "cross-sectoral" engagement.

Agency Needs vs. Site Managers' Records

Most camp managers did not have the background demographic data about site residents that agencies needed (*i.e.*, # of people per shelter, # of employed persons per shelter). The small TSST enumerator teams could not visit each and every shelter on a site visit in addition to assessing the condition of site-level infrastructure. Enumerators designed a short form (written in Sinhala or Tamil) that asked for the requested information. Upon reaching the site and with the help of site managers, the forms were distributed to every shelter. Occupants were instructed about how to complete the form. While the enumeration team inspectedthe site, shelter occupants completed the forms. At the end of the site visit, the enumerators collected the forms and helped occupants fill in missing information. Enumerators aggregated the data from the slips and input the data intothe PDA. This data-collection approach kept the research process on schedule.

Inability to Reach Conflict Affected Areas

The tsunami affected 11 of Sri Lanka's 25 districts. Due to the prevailing ethnic conflict at the time and the fact that the Liberation Tigers of Tamil Eelam (LTTE) controlled many tsunami-effected areas, the TSST team was able to assess site conditions and monitor the closure of sites in 8 of these 11 districts. Within two districts (Batticaola and Trincomalee), the team could not access all sites as some sites fell within LTTE-controlled sub-district units (Divisional Secretary Divisions). While the potential for a comprehensive and systematic island-wide database was possible, security concerns and the politics of working in conflict-affected areasforeclosed this opportunity. Despite lobbying by the UN and influential donors for the joint mechanism and its advantages, an island-wide assessment was not completed.

The Politics of the Resettlement Process

As the reconstruction process unfolded, themonitoring process highlighted ethnic tensions and issue of political patronage that historically have plagued Sri Lanka (Hyndman, 2011). The politics of aid and access to government resources is part of the post-disaster resettlement process. In a conflict-affected environment, the nature of aid distribution articulates with ethnic tensions and divides. Comparative analyses of 2006 and 2007 TSST data sparked numerous discussions about apparent differences in the resettlement process between the Sinhalese-dominated South and the Tamil and Muslim-dominated East as well as Government bias against the urban poor (Caron 2009 see map referenced in footnote 3). Evaluating the reconstruction process must take such local historical factors into account.

Conclusion and Recommendations

With 2015 as the International Year of Evaluation and in the lead-up to the 2016 World Humanitarian Summit, the lessons learned from Sri Lanka's TSST project implementation as highlighted in this chapter should encourage development practitioners and donors to:

1. Fund routine and systematic longitudinalhumanitarian assessments across all implementation phases (from response, relief and recovery through and into reconstruction, resettlement, and development);

2. Facilitate and support collaboration between agencies and governments to maximize efficient use of funds and to ensure that affect persons do not suffer research fatigue;

3. Invest in mainstreaming and institutionalizing the monitoring and evaluation process to preserve institutional knowledge,

4. Support flexibility and mid-course corrections as a result of new findings,

5. Sponsor regional coordination meetings among members of the humanitarian assessment community. Based on the TSST experience, members of the Sri Lankan project team traveled to Islamabad, Pakistan in 2008 to work with UN agencies there to adapt this approach, which later became the basis for the McRam assessments that took place in 2009-2010[1], and;

Recognize that politics and history influences how governments work with humanitarian agencies, how humanitarian agencies work with one another, and how governments and affected communities respond to and work with one another, and use evaluation data improve working relationships.

Appendix One

Total site-level vulnerability indicator (from UNICEF and UNOPS, 2006)

"The 12 indicators needed for calculating total vulnerability are" (ibid: 23):

PV   Population Vulnerability

SH   Shelter Vulnerability

 W   W Vulnerability

SN   Sanitation Vulnerability

AC   Access to Host Community

HC   Healthcare Vulnerability

PS   Availability of Public Services

FL   Flood Vulnerability

 R   Risk Vulnerability

LV   Livelihood Vulnerability

ED   Education Vulnerability

PR   Permanent Resettlement Vulnerability

---

4   For more on McRam and subsequent innovations to the TSST methodology, visit the Assessment Capabilities Project (ACAPS) website, http://www.acaps.org/

Weighting

Not every sector is equal in creating vulnerability. In order to express relative importance, "a weight is assigned to each sector, ranging from 1 to 9, the most important factor getting the highest value" (ibid: 23).

The calculated indicator value (see HC example in Appendix 2) is then multiplied by the following weight:

PV * 9 = PV weighed value

SH * 7 = SH weighed value

W * 9 = W weighed value

SN * 9 = SN weighed value

AC * 1= AC weighed value

HC * 8 = HC weighed value

PS * 2 = PS weighed value

FL * 6 = FL weighed value

R * 5 = R weighed value

LV * 5 = LV weighed value

ED * 3 = ED weighed value

PR * 4 = PR weighed value

To calculate total site-level vulnerability, all weighted values are added and divides by the sum of weights, in this case 68 (ibid: 24).

Total Vulnerability score: (PV+SH+W+SN+AC+HC+PS+FL+R+LV+ED+PR)/68.

Appendix Two

Health Indicator (from UNICEF and UNOPS, 2006)

"The elements needs for calculating the health care indicator: distance to natal care, distance to emergency obstetrics, # of visits to *(sic)* health staff [to] the site per month and the availability of medical supplies" (ibid: 15).

HC1: Distance from site to the nearest natal care facility

1 = < 3 km	3= 5-10 km
2= 3-5 km	4= > 10 km

HC2: Distance from site to the nearest emergency obstetrics facility

1 = < 3 km	3= 5-10 km
2= 3-5 km	4= > 10 km

HC3: Distance from site to the nearest hospital facility

1 = < 3 km	3= 5-10 km
2= 3-5 km	4= > 10 km

HC4: Whether or not a midwife visited the site in the past month

    2= yes   4= no

HC5: Whether or not a Public Health Inspector (PHI) visited the site in the past month

    2= yes   4= no

HC6: Whether or not mental health staff visited the site in the past month

    2= yes   4= no

HC7: Whether or not there is a first aid box on the site that is regularly refreshed with supplies

    1= yes, regularly refilled   4= no first aid box available

    2= yes, but not regularly refilled

References

Caron, C.M. 2009. "Left Behind: Post-tsunami Resettlement Experiences for Women and the Urban Poor in Colombo." In: Fernando, P., K. Fernando, and M. Kumarasiri (eds.) *Forced to Move: Involuntary Displacement and Resettlement – Policy and Practice.* Colombo: Centre for Poverty Analysis. pp. 177-206.

Government of Sri Lanka and Development Partners. 2005. *Sri Lanka: Post-tsunami Recovery and Reconstruction: Progress, Challenges, Way Forward.* Colombo: UNDP.

Hyndman, J. 2011. *Dual Disasters: Humanitarian Aid after the 2004 Tsunami.* Sterling, VA: Kumarian Press.

UNICEF and UNOPS. 2006. *Report Summary Guidelines.* Colombo.

Chapter 23

The 'GREEVIS' Framework for Evaluation of CSR Projects in India

Devanshu Chakravarti[1], Shailesh Nagar[2]
and Sirisha Indukuri[3]

N R Management Consultants India Private Limited (NRMC),
New Delhi
E-mail: [1]devanshu@nrmcindia.co.in; [2]snagar@nrmcindia.co.in;
[3]sirishai@nrmcindia.co.in

ABSTRACT

Corporate Social Responsibility (CSR) in India is transitioning from being a social philanthropy to being a professionally managed strategic endeavour.The Companies Act 2013 mandates every company, corporate, private limited or public limited with a net worth of ' 500 crores, or a turnover of ' 1000 crore, or net profit of ' 5 crore, to spend at least 2 percent of average net profit for the immediately preceding three financial years, on CSR activities. Interventions under CSR in India are highly varied and multi –sectoral in nature;ranging from stand-alone activities to those implemented in a project or programme mode and having short-term to long-term engagement in the targeted areas. Given this, it becomes difficult to use standard approaches for evaluation of CSR projects. This paper proposes a comprehensive framework evolved from action research,the GREEVIS framework,for assessment of CSR projects and programmes. The framework combinesparameters from the Development Assistance Committee (DAC) assessment criteria followed by Organisation for Economic Cooperation and Development (OECD) with parameters that are prioritiesof corporate houses in implementing CSR projects.

Keywords*:   Corporate social responsibility, Social philanthropy, Strategic endeavour, DAC, OECD, GREEVIS, Goodwill generation, Visibility.*

Corporate Social Responsibility in India: A Background

Over the past several decades, Corporate Social Responsibility(CSR) in India has graduated from "contributions and aid" to becoming "an essential part of the corporate stratagem" (Raju, 2014: page 3). It has evolved from a social philanthropy to a professionally managed strategic endeavour.

A professional approach was first adopted, possibly, by large conglomerates like the TATA group who formed various trustswith interventions based on a long-term development vision having a thematic or geographical focus. This new approach stemmed from the realization within the corporates that they needed to support development of the social-ecology within which they operated. Social movements and international agreements (like the Earth Charter,International Conventions on Child Rights and others) also pushed private sector organizations to include community development works within their business charters and project themselves as compliers to social and environmental issues.

Typology of CSR Initiatives

In India, the approach to CSR has been highly varied. The CSR projects can, either be strategic, well planned and outcome based; or, interventions may address immediate need, planned for one time or short duration and be output based.An analysis of data from 51 corporate foundations reveals that CSR activities are being undertaken across many sectors like Education, Health and Livelihood sectors (NGOBox, undated) (Table 23.1).

Table 23.1: Top Five Development Themes on which CSR Foundations are Working

Thematic Area	Education	Health	Livelihood	Women Empowerment	Environment
Number of CSR foundations working on it	39	33	10	9	8

Source: NGOBox, undated.

Within each sector, the interventions too were found to be variedin nature. For example, in the health sector, while Adani Foundation provides infrastructure support (like mobile health care units etc), Avantha Foundation helps augment government medical services at the grass-roots through village health workers (NGOBox, undated).A typology of CSR works is given in Table 23.2.

Need for a CSR Evaluation Framework

The Companies Act 2013mandates every company, corporate, private limited or public limited, which has a net worth of ' 500 crores or a turnover of ' 1000 crore or net profit of ' 5 crore, to spend at least 2 percent of its average net profit for the immediately preceding three financial years, on CSR activities. As per the provisions of the Companies (Corporate Social Responsibility Policy) Rules, CSR activities should not be undertaken in the normal course of business and must be done with respect to any of the activities mentioned in Schedule VII (as amended in February 2014) of The Companies Act 2013.

Table 23.2: Typology of CSR Initiatives

Strategic	Philanthropic
The projects/activities are strategic in nature addressing particular issues/problems/needs.	The projects/activities are short duration or one time.
The activities/projects are well planned.	The activities do not require much planning.
The activities/projects are outcome based.	The activities/projects are primarily output based.
The activities/projects require strategic partnerships with key stakeholders.	The activities/projects do not require strategic partnerships.
Examples could be watershed development, livelihoods etc.	Examples could be blood donation camps, providing desks and chairs in schools etc.

The implementation of the Act will infuse huge funds in the development sector in India. An assessment of net profit of the top 500 companies in India in the last three years suggests that an amount of ' 9600 crores would have to be spent by the private sector on CSR in the financial year 2014-15 (NGOBox, 2014).

The coming into effect of the Act on 1st April 2014, has led companies to relook at their existing CSR policy and assess the existing activities and projects. It is also incumbent on the companies to report and explain the mandated spending on CSR (Gupta, 2014). However, given the variety of works and approaches followed to implement CSR, coupled with the scale of activities, arriving at a standard approach for evaluation of CSR projectsis an area of challenge. Review of existing literature on the subject also reveals that "No standard method exists to approximate consequences of CSR on the community" (Raju 2014: page 5).

However, it is imperative that results of CSR activities are objectively validated for reporting of achievement as also for offering lessons for scaling up and inputs to future projects.

In this paper, the authorspropose a framework,evolved during the course of CSR project evaluations that was found most suitable for evaluation of CSR projects. The framework combines parameters from the Organisation for Economic Cooperation and Development (OECD) -Development Assistance Committee (DAC)assessment criteria, with parameters identified as CSR prioritiesby Indian corporates.

DAC-OECD Criteria: An Accepted Industry Standard for Assessment of Development Projects in India

For mid-term, end term and ex-post assessments of projects/programmes, while different outcome and output based frameworks are common, most evaluations prefer the results based OECD's DAC assessment criteria comprising the five parameters of Relevance, Effectiveness, Efficiency, Impact and Sustainability. The parameters can be adapted suitably for projects in any thematic domain with qualitative and quantitative indicators capturing project achievements under each parameter. While this framework can capture achievements under CSR projects, additional parameters defined by corporate prioritiesneed to be included in the framework for a comprehensive assessment of CSR project impacts.

Goodwill Generation and Visibility: Corporate Priorities for CSR Projects

For most companies operating in rural hinterlands, CSR projects are an important risk mitigation strategy,where activities are implemented to meet expectations and development priorities of the local communities. Hence, companies need to ensure that they maintain "durable contacts with society as a whole" (Raju, 2014: page 3) and generate goodwill. If 'Generating goodwill', is a major objective of CSR interventions, then,it is also an important assessment parameter for measuring the achievement of CSR projects.

Further, there is little information on CSR activities of corporates in the state or national media. Companies spending a substantial amount on rural/social development projects, want these facts to be widely known and get profiled as socially responsible organisations. Some companies undertake activities in public places like bus stands, religious places, etc. not only for greater goodwill generation but for greater visibility.Hence, visibility is also an important objective that corporates strive to achieve through CSR activities and it becomes an important assessment criterion for measuring achievement/impact of CSR projects.

The 'GREEVIS' Framework

While the key DAC-OECD parameters stay pertinent for evaluation of CSR projects, it is also pertinent to measure 'goodwill generation' and 'visibility gained'in these projects. A comprehensive framework, 'GREEVIS' with seven key evaluation parameters-**G**oodwill Generation, **R**elevance, **E**ffectiveness, **E**fficiency, **V**isibility, **I**mpact, and, **S**ustainability,was evolved from action researchby NR Management Consultants India Pvt. Ltd (NRMC)[1], while working on different CSR project evaluations.

The GREEVIS framework is detailed in Table 23.3.

Compared to the operational scale of Government, bilateral, multilateral or even International NGOs, CSR projects are quite small. The revised CSR guidelines for Central Public Sector Enterprises (2013) under Section 1.8.5 (page 30), states that, "It is recognised that small scale activities/projects under CSR and sustainability agenda of a company cannot be expected to have any significant social or economic or environmental impact, which can be easily measurable." Hence, the scale of most CSR projects limits the use of only limited quasi- experimental or non-experimental methodologies. The GREEVIS framework offers the flexibility of using these methodologies for activity specific or thematic evaluationusing both qualitative and quantitative indicators.

Conclusion

Several factors may act as constraints for effective application of the GREEVIS framework.

Section 1.2.6 (pp 5-6) of the revised CSR guidelines for Central Public Sector Enterprises (2013), states the requirement for "credible evidence of having made a fairly accurate assessment of the needs of the stakeholders, which would also help in

Table 23.3: GREEVIS Framework

The GREEVIS Framework for Assessment of CSR Projects and Programmes

Criteria	Description and Key Questions to be Addressed
Goodwill Generation	The extent to which the programme/project activities have evoked a sense of warmth, trust and friendliness towards the company from the local communities and other stakeholders.
	☆ At which levels has Goodwill been generated - Local, State and National?
	☆ How have the perceptions and attitudes regarding the company changed among the target groups, post the initiation of the project?
	☆ Is there confidence on the sustainability of the project intervention?
	☆ What is the level of cooperation among the target groups, post the project intervention?
Relevance	The extent to which the CSR activities are suited to the priorities of the target group.
	☆ To what extent are the objectives of the programme/project valid in view of the field context?
	☆ Are the activities and outputs of the programme/project consistent with the overall goal and the attainment of its CSR objectives and the CSR vision of the company?
	☆ Are the activities and outputs of the programme/project consistent with the categories in Schedule VII (as amended in February 2014) of Companies Act 2013?
Effectiveness	The extent to which the programme/project activities attain their objectives.
	☆ To what extent were the objectives achieved?
	☆ What were the major factors influencing the achievement or non-achievement of the objectives?
Efficiency	A measure of outputs (qualitative and quantitative) in relation to the inputs; measure of progress made with respect to what was planned.
	☆ Were activities cost-efficient?
	☆ Were objectives achieved on time?˙Was the programme or project implemented in the most efficient way compared to alternatives?
Visibility	Measures how 'visible' were the CSR activities and the recall among target groups.
	☆ At what level was visibility planned- local, state and national?
	☆ How appropriate was the targeting?
	☆ How appropriate was the message?
	☆ Was the recall among the target group as per expectations?

Contd...

Table 23.3—Contd...

The GREEVIS Framework for Assessment of CSR Projects and Programmes

Criteria	Description and Key Questions to be Addressed
Impact	The positive and negative changes produced by the CSR activities, directly or indirectly, intended or unintended on the social, economic, or environmental indicators. ☆ What has happened as a result of the programme or project? ☆ What real difference has the activity made to the beneficiaries? ☆ How many people have been affected?
Sustainability	Measures whether the benefits of the activities are likely to continue after funding has been withdrawn. Programmes/Projects need to be environmentally as well as financially sustainable. ☆ To what extent did the benefits of programme/project activities continue after funding was stopped? ☆ To what extent was the maintenance of the assets taken up by the target audience? ☆ What were the major factors which influenced the achievement or non-achievement of sustainability of the programme or project?

Source. Adapted from DAC-Criteria; discussions with stakeholders.

Figure 23.1: GREEVIS Framework for CSR Assessments.

making a fair estimation of the social or environmental impact after the conclusion of the activity."However, most CSR projects lack adetailed need assessmentcompleted prior to the project implementation.

Most CSR projects focus largely on addressing 'practical needs' like infrastructure creation with disproportionatelyless spending on addressing 'strategic needs' like trainings and capacity building. Although such investment helps in creatingthe much needed facilities, without adequate sensitization, sustainability issues are not addressed adequately and community ownership remains a challenge. And this constrains results under the sustainability parameter.

Thirdly, most CSR projects do not attempt to influence government policies based on their learning and successful experiences. Documentation initiatives are limited to showcasing of activities and never intended for policy advocacy at the state or national levels. The measurement of policy impactis hence largely subdued in terms of investments made in CSR projects.

Despite these constraints, the GREEVIS framework offers the most comprehensive evaluation framework for the current nature of CSR projects.The framework is evolving and which will get enriched from learnings from future field application.

Acknowledgements

The authors thank corporate representatives, communities and other stakeholders, interactions with whom proved useful in the evolution and field testing of the framework.

References

Department of Public Enterprises, Government of India, 2013. Guidelines on Corporate Social Responsibility and Sustainability for Central Public Sector Enterprises. New Delhi: Department of Public Enterprises, Government of India.

Department of Public Enterprises, Government of India, 2014. Guidelines on Corporate Social Responsibility and Sustainability for Central Public Sector Enterprises. Department of Public Enterprises, Government of India.

Gupta, Ananda Das, 2014. Implementing Corporate Social Responsibility in India: Issues and Beyond in Ray, Subhasis, and Raju, S. Siva (eds.), 2014. Implementing Corporate Social Responsibility: Indian Perspectives. New Delhi: Springer

Ministry of Corporate Affairs, 2014. Notification: Amendment to Schedule VII, Companies Act 2013. New Delhi: Government of India

Ministry of Law and Justice, Government of India, 2013. The Companies Act, 2013. New Delhi: Ministry of Law and Justice, Government of India.

NGOBox, 2014. CSR Research Series: Mandatory CSR Spending for Big 500 Companies (for FY 2014-2015). New Delhi: NGOBox

NGOBox, undated. 51 CSR Foundations (Corporate) in India. New Delhi: NGOBox

OECD, 1991. The DAC Principles for the Evaluation of Development Assistance. Paris: OECD. October 27, 2014 – Online. Available at: http://www.oecd.org/development/evaluation/daccriteriaforevaluatingdevelopmentassistance.htm

Raju, S. Siva, 2014. Measuring Performance of Corporate Social Initiatives: Some Methodological Issues. in Ray, Subhasis, and Raju, S. Siva (eds.), 2014. Implementing Corporate Social Responsibility: Indian Perspectives. New Delhi: Springer.

Ray, Subhasis, and Raju, S. Siva (eds.), 2014. Implementing Corporate Social Responsibility: Indian Perspectives. New Delhi: Springer.

Chapter 24

Use of Econometric Techniques in Programme Evaluation

P.K. Anand

Senior Consultant,
NITI Aayog, New Delhi
E-mail: pkanand1@yahoo.com

ABSTRACT

In India 66 Centrally Sponsored Schemes, cover a wide spectrum of Socio-Economic interventions. In order to evaluate impact of these Schemes use of econometric analysis, can facilitate in modification of Schemes to enhance their outcomes. This paper covers some case studies from the angle of evaluation from India (on Public Distribution, Sanitation and Health), and from abroad (Sri Lanka on Poverty and Kenya on Agriculture), to touch upon some of the challenges faced and successfully handled using econometrics.

Objective of Econometric Analysis

The core objective of any econometric analysis is to suggest modifications/ corrective actions of a policy to maximize marginal utility of limited appropriate manpower, materials and financial resources and to build these resources for sustained growth, welfare etc., with cooperation of various stakeholders in response to convincing econometric evidence.

Econometrics helps in this direction, as most such relationships are probabilistic and not deterministic. To achieve this objective, Gujarati (2005) crystallizes that an econometric analysis proceeds along the logical lines showing in Figure 24.1:

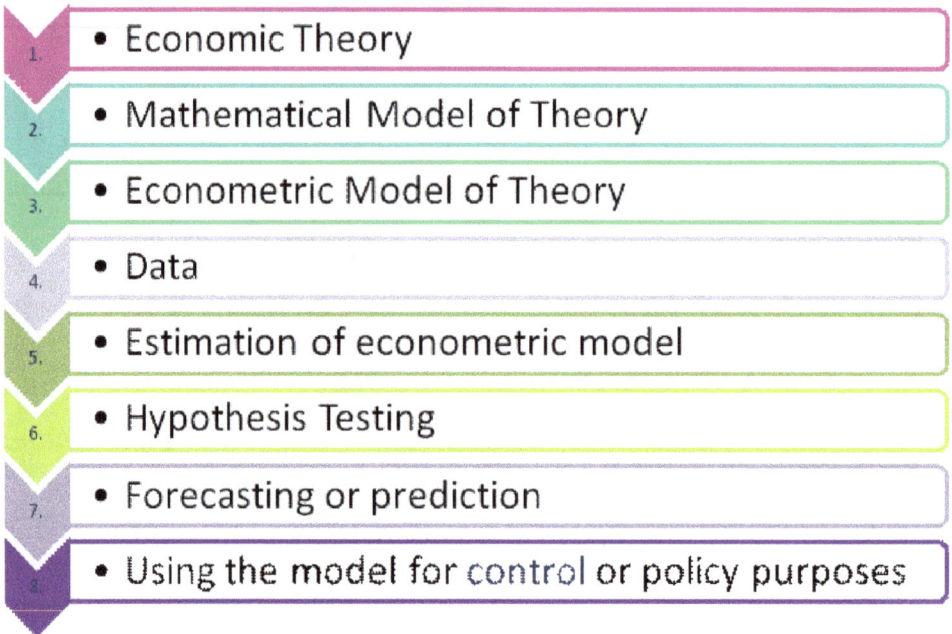

1. • Economic Theory

2. • Mathematical Model of Theory

3. • Econometric Model of Theory

4. • Data

5. • Estimation of econometric model

6. • Hypothesis Testing

7. • Forecasting or prediction

8. • Using the model for control or policy purposes

Figure 24.1

Choice of Variables

For evaluation of government sponsored programmes, broad categories of variables of interest vary from study to study, but some frequently used groups are socio-economic, demographic and programme related variables. For application of econometric techniques measurement of (a few to most) variables facilitates in capturing relationships. Still as the explanatory power of tests applied reveals, all changes are not fully explained (e.g. Adjusted R^2 in OLS regression is rarely close to 1), for which one of the reasons could be omitted variable bias. Or sometimes, a valid looking variable may need to be dropped due to the problem of multicollinearity, say education level of father when education level of mother (which generally explains better) is already being used as an explanatory variable, and both impacts can't be disentangled (Maddala 1977).

Task before an econometrician is to assess relationships and marginal effects (change in Y when X_1 changes by one unit) in the application of regression techniques, as this sets it apart from computation of correlations. Causality plays an important role in choice of variables, as once causality runs from X_1 to Y one can set Y as the dependent (or response) variable and X_1 (alongwith others like X_2, X_3, X_4 ...) as an independent / explanatory/ predictor variable. For example to explain the periodic number of malaria caused deaths, the explanatory variables may be availability of health centres (public and private), distance of such centres, availability of medicines, awareness building campaigns, cleanliness related and other preventive steps and so on. Still one of the omitted variables could be the negative externality of stagnant water in small/ medium/ major irrigation water bodies.

In real life situations, causality may not be flowing in an obvious manner. There can be such pair of variables between which **causality may run both ways**, and this may necessitate say, **Granger causality tests** (say, on time series lagged data) to ascertain the type and direction and relationship. Many relationships between income and education or between money supply and GDP falling in this category may require a word of caution to handle.

Coming to the predictive power of a model, one should remember **Lucas critique (1976)**, that it is naive to try to predict the effects of a change in economic policy entirely on the basis of relationships observed in historical data, especially highly aggregated historical data. This could be for not taking into account the degree to which estimated functional forms fail to be deep. For instance, if old age/ widow/ disability pension schemes are introduced afresh or levels of assistance are remarkably enhanced (say, if topped up by a State in India), latter period household (real) expenditure data needs to be taken with a pinch of salt.

More light can be thrown through the studies next covered.

India : Targeted Public Distribution System

This 2005 study (no. 189) was conducted by the Programme Evaluation Organisation (PEO) under erstwhile Planning Commission from a sample of 3,600 households from 60 districts across 18 states. It logically categorises supply and demand side variables.

The OLS results for first regression can be expressed as:

%PD = 95.04 +0.116INS −0.362OWN −0.142ASTS −0.165PRE +0.051PR − 66.11D

(12.85)　(1.91)　　　(-1.67)　　　(-2.08)　　(-2.42)　　(1.68)　　(-6.09)

(Figures in brackets are t values). The R^2 value comes to 0.62.

Notably, all variables were averaged for sample households at the district level.

It had precisely set 'Percentage of BPL respondents lift Public Distribution System grains (per cent PD)', as the dependent variable.

The explanatory variables (all for BPL respondents) covered were as follows,

INS = Per cent of respondents allowed to buy in instalments

OWN = Per cent of rice and wheat requirements met from own production / wages in kind.

ASTS = Per cent respondents possessing specified assets

PRE = Per cent respondents preferring local variety to PDS variety

PR = The ratio between the weighted average of market price of rice and wheat and the weighted average PDS price.

D = A dummy variable for the two outlier observations.

Another OLS equation evolved; with a changed model specification was follows:

Qpd = −0.79 +1.037ENT +0.020INS −0.067OWN +0.003ASTS −0.002PR −0.031FS

(-0.023)　(6.91)　　(1.99)　　(-1.59)　　　(0.189)　　(0.30)　　(-0.54)

(Figures in brackets are t-stat values). R^2 is 0.61 for this model.

Qpd = Average PDS lifting of cereals by respondents reporting lifting

ENT = Average district-wise entitlement, not adjusted for supply side shortages.

FS = Ratio of number of households with size less than or equal to 2 in total.

(All othear variables are as already explained.)

Conclusions

It was found that INS and PR affect the decision to buy from PDS (per cent PD) positively, ASTS, OWN and PRE affect it negatively. Since the dependent and the explanatory variables (except the dummy) in the first regression are given in percentages, the regression coefficients also serve as elasticities and are readily amenable to policy conclusions.

The second regression results strongly suggest that once the decision to lift food grains is made, the quantity to be lifted is predominantly determined by the supply side factors.

Taken together ASTS and PR reveal once the asset owner decides to buy from PDS, he buys as much as the poorer BPL cardholder does.

India : Total Sanitation Campaign

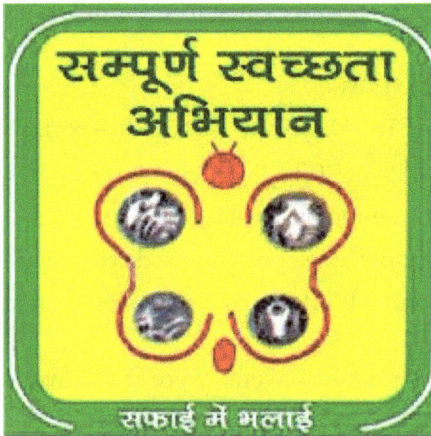

सम्पूर्ण स्वच्छता अभियान
सफाई में भलाई

ग्रामपंचायत निजामपुर ता-नवापुर
डेमो वैयक्तिक शौचालय बांधकामाची प्रगती

This 2013 evaluation study (no. 219) on Total Sanitation Campaign (TSC) was conducted by the Programme Evaluation Organisation (PEO) under erstwhile Planning Commission from a sample of 11,452 households across 20 States.

20 States	122 Districts	1207 Gram Panchayats	11,452 Households

Objectives of the Study

The objectives of the study included to ascertain the impact of the ongoing TSC programme (now known as 'Nirmal Bharat Abhiyan'), of which Open Defecation is covered in this Paper.

Variables Used

Notably, all the variables used were categorical variables, with the exception of 'number of grassroot workers recruited by Gram Panchayat,' which was a cardinal variable. The dependent variable used was related to Open defecation (OD), and assigned binary values as 0 for Open Defecation and 1 for No Open Defecation. Thus the desired policy variable was given a positive number, so that the significant explanatory variables appearing with positive sign can at a glance be termed as desirable to push the policy.

The explanatory variables used, which were found significant, are as follows: Family Size upto 3 members (1), 3 to 5 (2) and above 5 (3); Education level, "proximate illiteracy" of highest level of a member Illiterates (1), Primary, Upper Primary, High School and Unschooled literates (2), and above this level (3). Number of grass root workers in GP was the only scale variable.

Adequate toilet facility (1), Adequate Water supply (1), Awareness of TSC (1), Awareness of water-borne diseases (if improper sanitation) (1), BPLIncen–interaction term between BPL and Incentive (1), Presence of Village Water and Sanitation Committee in the GP (1), Grievance Redressal Mechanism (1), PRI Role in monitoring and funding (1), were dummy variables.

Table 24.1: Total Sanitation Campaign : Dependent Variable
No Open Defecation (1) (0 being for Open Defecation)

	B	S.E.	Wald	df	Sig.	Exp(B)
Fam_Sz			81.267	2	0.000	
Fam_Sz(1)	0.937	0.109	73.885	1	0.000	2.552
Fam_Sz(2)	0.408	0.070	34.377	1	0.000	1.504
Bpl(1)	0.155	0.112	1.919	1	0.166	1.167
adeq_toi(1)	1.110	0.085	170.783	1	0.000	3.035
enuf_wat(1)	0.431	0.069	38.973	1	0.000	1.539
awareDis(1)	0.346	0.090	14.868	1	0.000	1.413
aware_TSC(1)	1.134	0.103	121.745	1	0.000	3.109
HH_Educ			86.176	2	0.000	
HH_Educ(1)	−1.266	0.137	85.320	1	0.000	0.282
HH_Educ(2)	−0.423	0.080	27.819	1	0.000	0.655
BPLIncen(1)	−1.310	0.259	25.644	1	0.000	0.270
wsc_formed(1)	0.864	0.064	183.689	1	0.000	2.373
Griev_Red(1)	0.611	0.087	49.792	1	0.000	1.842
PRI_Role(1)	0.193	0.074	6.775	1	0.009	1.213
grassrrot_worker	0.043	0.005	64.563	1	0.000	1.044
Constant	−0.671	0.272	6.074	1	0.014	0.5110

Results

As is seen from the above logit regression results, smaller family size, adequate toilet provision, enough water in toilets, awareness of water-borne diseases, awareness of TSC, higher education level, grassroot workers, Water Sanitation Committee, Grievance Redressal raised odds ratio in favour of no open defecation. The way BPL and Incentive were defined, BPLIncen implied relatively BPL/ availing incentive having a favourable odds ratio (1/0.270 = 3.706) in favour of no open defecation.

Conclusions

Equipped with logit results, the study culls out key policy variables that favourably affect the objectives of the ongoing Total Sanitation Campaign (TSC), rechristened Nirmal Bharat Abhiyan. Notably, it also underscores other key variables, such as education, family size and water availability that fall outside the scope of TSC but also have a significant bearing on the success of the TSC, and stresses that mere provisioning of toilet facilities is not enough.

Sri Lanka : Poverty Reduction

A study by Ranathunga (2000) on household poverty in Sri Lanka used OLS and Probit regressions to analyse factors explaining poverty. OLS is applied with 'log expenditure per capita per month' as the dependent variable for a 76,442 observations set. The study rightly selects (ln) consumption expenditure instead of income as the dependent variable arguing that income data in any country is known to be less reliable than the consumption data in household surveys, since income data is often under-reported and there are difficulties of quantifying (eg. self-employment and capital income) and time variable also influences it (due to seasonality of earnings).

OLS Results

R^2 value of OLS estimates comes to 0.35. Explanatory variables found significant at 1 per cent level that enhance household expenditure include education of household head, highest education level in family, spouse education, both spouses working in government/ semi government, foreign remittance, local remittance; and that diminish household expenditure include household size, dependency ratio (number of household members below 15 and above 60 years of age), both spouses in private sector/ both self-employed etc.

Interestingly, female-headed households are less likely to be in poor in rural sector but more likely in urban sector. Age of the head of the household was found insignificant. Higher number of children increased negative effects. While the families' with less than two children are less likely to be poor in rural sector, it is an insignificant variable for the urban and estate sectors.

Probit Analysis

For probit analysis, the binary dependent variable was 'Household under the poverty line' or 'Otherwise' linked to household expenditure and results were human capital and remittances driven.

India : Health Sector

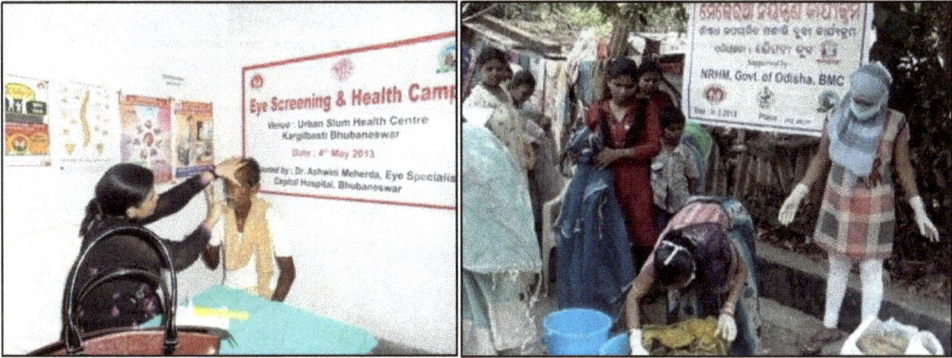

This 2012 evaluation study (no. 217) conducted by the Programme Evaluation Organisation (PEO) under erstwhile Planning Commission, through Population Research Centre, Institute for Economic Growth, drew a sample from 37 districts of 7 states was covered, out of 7,400 rural households covered, data on Utilization of Antenatal Care was elicited from 1,584 pregnant women.

Variables and Results

The multinomial logit regression technique was utilized to highlight the net effects of explanatory (predictor) variables on utilization of public or private health care facilities etc.

Here, the dependent variables of interest emanate from are (i) use of ANC from public medical sector (P_1) (ii) use of ANC from private medical sector (P_2) and (iii) no use of ANC care (P_0). The regression is run on log of ratios (P_1 / P_0) and (P_2 / P_0), where the sum of three probabilities $P_0 + P_1 + P_2 = 1$ and its results are as follows:

Table 24.2: Multinomial Logit Regression Coefficients of Rural Antenatal Care Utilization (ANC)), withNo-Use as Reference Category

Predictor Variables	ANC Govt. Log (P1/P0)		ANC Pvt. Log (P2/P0)	
	Coeff. (βi)	(μi)	Coeff. (βi)	(μi)
Intercept	1.281	**0.01**	1.669	0.11
WEDN (Women's Education)	0.135	**0.02**	0.225	**0.04**
PEI (Per Earner Income of the Household)	0.193	**0.00**	0.069	0.63
TOHBD (Type of House–ownership[1] binary dummy)	0.213	0.21	0.189	0.60
SKF (Separate kitchen facility)	0.664	**0.00**	1.241	**0.00**
STF (Separate toilet facility binary)	0.071	0.66	0.069	0.84
PDWFBD (Potable drinking water facility source)	−0.624	**0.00**	0.645	0.06

Contd...

Table 24.2–Contd...

Predictor Variables	ANC Govt. Log (P1/P0)		ANC Pvt. Log (P2/P0)	
	Coeff. (βi)	(µi)	Coeff. (βi)	(µi)
FUCBD (Fuel used for cooking)	−0.638	**0.01**	−1.073	**0.02**
ASHAVBDTN (ASHA's visits to household)	0.779	**0.00**	0.071	0.84
ASHAMBDTN (ASHA's carrying and distribution of free medicines)	0.595	**0.00**	−0.576	0.12
ASHASBDTN (ASHA's sensitizing/counselling with women on sanitation and obstetric care)	0.704	**0.00**	−0.089	0.82
VHNDBD (Holding Village Health and Nutrition Day)	−0.088	0.58	−0.204	0.56
VHSCBD (Village Health and sanitation Committee's role)	0.135	0.40	−0.228	0.54
DPHC (Distance of first referral unit, like PHC)	−0.202	**0.00**	0.276	**0.05**

n = 1584; Suffix BD used in various variables is Binary Dummy.

(µi): Level of Significance.

Conclusions

Positive effect of women's education on utilization from private was higher compared to public facilities.

Household's per earner income was positive for availing from public health but insignificant from private facilities.

Similarly, other household economic status variables like ownership of the house, having separate kitchen and separate toilet facility, cleaner water source, cleaner cooking fuel in the house are favourable for availing public facilities.

ASHA's role of visits, medicines distribution and counselling was positive on availing public facilities.

Holding Village Health and Nutrition Day as well as role of Village Health and Sanitation Committees didn't depict impact in motivating for public facilities.

Distance of first referral unit (District Hospital/CHC) from the village had negative impact on availing public health facilities, but interestingly led to higher recourse to private facilities.

Kenya : Agriculture Sector

This study of 2012 (Ayuva *et al.*) covers a sample of 150 farmers from Bungoma County, which is one of the areas in Kenya where maize is produced on small-scale basis. However, the County is facing soil nutrient depletion due to continuous and unsustainable cultivation of land. Various interventions have sensitized farmers into adopting Organic Soil Management Practices (OSMP) of enhancing soil fertility and upholding environmental sustainability. The study was aimed at establishing the most preferred OSMP among farmers and the factors influencing the choice of these techniques. Notably, soil erosion was less on the farms that adopted an OSMP. Thus OSMP facilitated sustainable maize production, averting soil nutrient mining.

Dependent Variables

Accordingly, as the choices of Dependent variable could be one of the 6 most preferred OSMPs or 'No adoption', multinomial Logit regression was done on the following 7 variables choices, with percentage of most preferred choices manifested in the Chart:

1. Use of Farm Yard Manure
2. Use of compost manure
3. Agro Forestry
4. Crop rotation
5. Planting Leguminous crops
6. Incorporation of crop residues or
7. No adoption (out of the preceding i to vi choices)

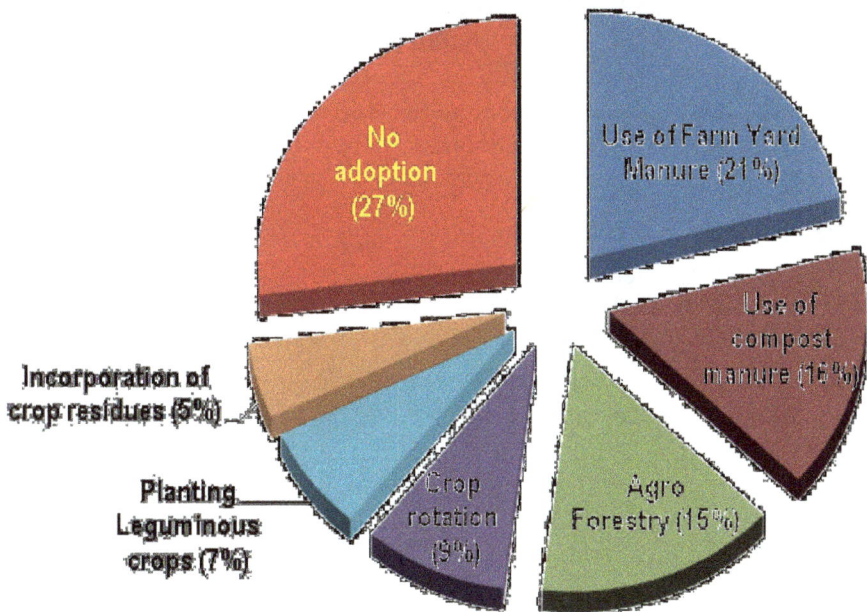

Explanatory Variables

The model tested for the 7 dependent, and 15 explanatory variables namely number of years of formal education of the household head, age of the chief decision maker, her/his gender, household size, farm size, farming experience, off-farm income, land tenure (ownership), credit accessed, extension contacts access, training sessions attended, erosion slope, member of a group, farm distance from homestead and perception towards organic farming.

Policy Recommendations

The study recommended that policies in support of OSMP should disaggregate farmers according to their socioeconomic, farmer and farm characteristics and suggested targets as:

1. For **crop rotation** - with large land holdings and relatively less slope bearing land.
2. For **crop residues** - older with small household size and to support them with more extension contacts.
3. For **farm yard manure** - young women headed and to encourage to form groups.
4. For **compost manure** - with small, less slope bearing land holdings to whom frequent extension contacts be provided.
5. For **Agro-forestry** - male headed with larger land holdings, large household membership and better security of tenure for land.

The study does not give any specific prescription to promote leguminous crops, as significant explanatory variables for it have small marginal effects. This remarkable study does full justice to MNL regression technique by prescribing dependent variable wise policy prescriptions as above.

In general, one can conclude from above studies that for evaluation of any programme, relevant econometric techniques can be applied and results utilised as inputs for appropriate policy modifications.

References

1. Ayuya, Oscar I.; Kenneth, Waluse S. and Eric, Gido O.; (July 2012). 'Multinomial Logit Analysis of Small-Scale Farmers' Choice of Organic Soil Management Practices in Bungoma County, Kenya', Current Research Journal of Social Sciences 4(4), pp 314-322.

2. Koutsoyiannis, A., *'Theory of Econometrics'*, McMillan Press Ltd., Hampshire, second edition 1977.

3. Lucas, Robert E., Jr. (1976). 'Econometric Policy Evaluation: A Critique" in: K. Brunner and A. Meltzer (eds.), The Phillips Curve and Labor Markets, Carnegie-Rochester Conference Series on Public Policy, Volume 1, pages 19-46.

4. Maddala, G.S., (1977). 'Econometrics', McGraw_Hill Book Company, International Edition, pp 183.

5. Planning Commission, (2013). 'Twelfth Five Year Plan 2012-2017 – Faster, More Inclusive and Sustainable Growth', New Delhi (Three volumes).

6. Programme Evaluation Organisation, Planning Commission, (2005). 'Performance Evaluation of Targeted Public Distribution System', (no. 189) New Delhi.

7. Programme Evaluation Organisation, Planning Commission, (2011). 'Evaluation Study on National Rural health Mission (NRHM) in Seven States', (no. 217) New Delhi.

8. Programme Evaluation Organisation, Planning Commission, (2013). 'Evaluation Study on Total sanitation Campaign', (no. 219) New Delhi.

9. Ranathunga, Seetha P.B., (August 2010). 'The determinants of household poverty in Sri Lanka: 2006/2007', Department of Economics, University of Waika, Munich Personal RePEc Archive (Paper 34174).

www.ingramcontent.com/pod-product-compliance
Lightning Source LLC
Chambersburg PA
CBHW050515190326
41458CB00005B/1546